Mock Papers for MRCPI part I

Three Mock Tests With 300 BOFs

Second Edition

This page was intentionally left blank

Second Edition

Mock Papers for MRCPI Part I

Three Mock Tests With 300 BOFs

Osama S. M. Amin
MD, FRCP(Edin), FRCP(Glasg), FRCPI, FRCP(Lond), FACP, FAHA, FCCP

Clinical Associate Professor in Neurology
International Medical University,
Kuala Lumpur, Malaysia

Osama S. M. Amin
Clinical School, International Medical University,
Jalan Rasah, 70300 Seremban, Negeri Sembilan,
Malaysia
Email: dr.osama.amin@gmail.com, osamashukir@imu.edu.my

First Edition: 2009.
Second Edition: 2016
ISBN: 978-1-365-58515-9

Disclaimer:
This book was written depending on reliable sources. However, while every effort has been made to ensure its accuracy, no responsibility for loss, damage, or injury occasioned on any person acting or refraining from action as a result of information contained herein can be accepted by the author or publisher.

Distributed by Lulu Press, Inc. Northern Carolina, USA.

Dedication

To my lovely wife Sarah, daughter Awan, and baby girl Naz. Without their kind help and patience, this books would not have been published.

Table of Contents

Acknowledgements

I woul like to thank my dear patients; their real clinical scenarios were used to formualte and generate these questions.

A special gratitude goes to my other half, wife Sarah, for her endless support and encouragement, and of course her extreme patience.

Osama

Preface

"The very first step towards success in any occupation is to become interested in it",
Sir William Osler (1849-1919).

In this book, you will find 3 mock papers. Each one contains 100 questions in a Best of Five (BOF) format. Self-assess, try to complete each paper within 3 hours, check out your answers, read the explanation, and re-read accredited medicine textbooks to fill in the gap in your knowledge. Each question has an *"objective"*; try to review what the objective is about. This is the only self-assessment book specifically written to imitate the MRCPI part I examination. Each mock paper in the first edition of this book contained 50 BOFs and 50 MCQs formats.

The Royal College of Physicians of Ireland states that the MRCPI General Medicine Part I examination consists of one paper with 100 Single Best Answer (i.e., best of five) questions. You have three hours to complete the exam. There is no negative marking and each question is equally weighted. Single Best Answer questions consist of a "vignette" (clinical scenario) followed by five possible answers. You must select the single best answer.

Questions in Part I are selected to achieve a balanced spread across medical specialties, the sciences underlying evidence-based medical practice, and the basic skills required in general medicine. Questions will be on common or important diseases in hospital or community practice as outlined in the curriculum for Basic Specialist Training in General Internal Medicine. At least 75% of the questions will concern direct clinical care of inpatients and outpatients in hospital medical practice. Correct answers will be within up-to-date guidelines for diagnosis and management. Basic science questions may cover anatomy, bacteriology, biochemistry, ethics, genetics, immunology, metabolics, physiology, principles of evidence-based practice and statistics.

In writing this book, I have tried to cover the most commonly encountered examination themes. Remember, this is just a mock test. For a more comprehensive self-assessment, try my "Get Through MRCP part 1", Self-Assessment for MRCP(UK) part 1", and "Neurology: Self-Assessment for MRCP(UK) and MRCP(I)" books.

Good luck with your career and exams!

Osama S. M. Amin
December 2016

References and Recommended Readings

1. Walker B, Colledge NR, Ralston S, Penmen I. *Davidson's Principles and Practice of Medicine, 26th edition*. London: Churchill Livingstone; 2014.

2. Kumar P, Clark ML. *Kumar and Clark's Clinical Medicine, 8th edition*. Philadelphia: Saunders Ltd.; 2012.

3. Longo D, Fauci A, Kasper D, Hauser S, Jameson J, Loscalzo J. *Harrison's principles of Internal Medicine, 18th edition*. New York: McGraw-Hill Professional, 2011.

4. Dale DC, Federman DD, *ACP Medicine, 3rd edition*. Philadelphia: BC Decker Inc.; 2007.

5. Barrett KA, Barman SM, Boitano S, Brooks H. *Ganong's Review of Medical Physiology, 24th edition*. New York: McGraw-Hill Education / Medical; 2012.

6. Lang TA, Secic M (eds.). *How to Report Statistics in Medicine, Annotated Guidelines for Authors, Editors, and Reviewers, 2nd edition*. Philadelphia: The American College of Physicians, 2006.

7. Rosene-Montella K, Keely EJ, Lee RV, Barbour LA (eds.). *Medical Care of the Pregnant Patient, 2nd edition*. Philadelphia: The American College of Physicians; 2007.

This page was intentionally left blank

Mock Paper Number One

100 Best of Five Questions

This page was intentionally left blank

1) A 34-year-old woman was diagnosed with relapsing-remitting multiple sclerosis before 1 year. Today, she presented with an attack of transverse myelitis. You examined her and found several neurological sings. You tried to remember the localization of these neurological deficits within the neuraxis. Which one of the following statements with respect to localization in clinical neurology is *correct*?

   a. Aphasia is a posterior fossa sign
   b. Apraxia indicates a spinal cord pathology
   c. Paraparesis can be due to cerebral hemispheric lesions
   d. Brown-Séquard syndrome is seen in midbrain lesions
   e. Locked-in syndrome can result from lesions affecting the upper midbrain

2) A 60-year-old man was brought to the Emergency Department. He had been experiencing severe substernal chest pain over the past 2 hours. You considered unstable angina. All of the following are "high risk factors" for unstable angina, *except*?

   a. Raised serum troponin-I level
   b. Clinical features of heart failure
   c. Transient ST-segment elevation
   d. No past history of acute coronary syndrome
   e. Post-infarct angina

3) A 31-year-old woman presented with a 4-month-hisotry of anxiety, palpitation, sweating, and menstrual disturbances. You diagnosed thyrotoxicosis. All of the following can result in hyperthyroidism with "negligible radioiodine uptake test", *except*?

   a. Graves' disease
   b. Thyrotoxicosis factitia
   c. Struma ovarii
   d. Post-partum thyrotoxicosis
   e. Amiodarone-induced thyrotoxicosis

4) A 54-year-old man had been experiencing progressive dysphagia over the past 3 months. He had unintentional weight loss and anemia. You did esophagoscopy and found esophageal cancer. Which one of the following is *not* a recognized etiological factor for the development of squamous cell carcinoma of the esophagus?

   a. Achalasia of the cardia
   b. Celiac disease
   c. Tylosis
   d. The presence of post-caustic stricture
   e. Gastroesophageal reflux disease

5) A 47-year-old woman presented with pallor, malaise, diffuse ecchymosis, and fever. Complete blood counts and blood films examination of the patient revealed acute lymphoblastic leukemia. Which one of the following medications is commonly used in the "remission induction" phase of acute lymphoblastic leukemia?

   a. Intravenous daunorubicin
   b. Intravenous mitoxantrone
   c. Intrathecal vincristine
   d. Oral thioguanine
   e. Oral methotrexate

6) A 64-year-year-old man has been diagnosed with Wegener's granulomatosis. Which one of the following is *not* a feature of this form of vasculitis?

   a. Enophthalmos
   b. Oral ulcers
   c. Pleural effusion
   d. Hearing loss
   e. Rhinitis

7) A 30-year-old man presents with unexplained and persistent fever. He is homosexual. You run a battery of investigations and find that he is HIV-positive. Which one of the following may be the cause of this unexplained fever?

   a. Disseminated CMV infection
   b. Kaposi sarcoma

c. Cryptosporidium small bowel infection
d. Oral thrush
e. HIV nephropathy

8) A 69-year-old man was admitted to the coronary care unit because of chest pain and palpitation. He was diagnosed with 3-vessel coronary artery disease last year. Which one of the following anti-arrhythmic medications predominantly reduces the maximum velocity of the upstroke of action potential?

a. Flecainide
b. Amiodarone
c. Digoxin
d. Sotalol
e. Verapamil

9) A 43-year-old man was referred to the Emergency Department from a rural hospital. His blood urea and serum creatinine are high. Which one of the following is *not* an intrinsic cause of acute renal failure?

a. Iodinated contrast agents
b. Intravenous gentamycin
c. Protracted vomiting
d. Malignant hypertension
e. Myoglobinuria

10) While investigating a 54-year-old woman, you found that her serum amylase is elevated. You further investigated her pancreas, and all investigations turned out to be normal. All of the following are non-pancreatic causes of hyperamylasemia, *except?*

a. Acute viral parotitis
b. Ruptured ectopic pregnancy
c. Following ERCP
d. Lactic acidosis
e. Renal insufficiency

11) A 34-year-old man was admitted to the intensive care unit after ingesting a large amount of oral paracetamol. His liver enzymes are very high. Which one of the following is *not* a complications of fulminant hepatic failure?

a. Hypoglycemia
b. Thrombocytosis
c. Encephalopathy
d. Agitation
e. Cerebral edema

12) A 45-year-old alcoholic man was found to have low serum level of phosphate. Which one of the following is *not* a consequence of severe hypophosphatemia?

a. Rhabdomyolysis
b. Polycythemia
c. Osteomalacia
d. Heart failure
e. Confusion

13) A 29-year-old man visited the Emergency Department because of severe headache. You took a thorough history and examined the patient thereafter. Your preliminary diagnosis was cluster headache. Which one of the following is a characteristic feature of cluster headache?

a. Nasal dryness
b. Forehead sweating
c. Lid retraction and eyelid edema
d. Pupillary dilatation
e. Occipital location

14) A 79-year-old man underwent several laboratory tests for Paget's disease of the bone. Which one of the following is *consistent* with that diagnosis?

a. Raised serum alkaline phosphatase
b. Hypocalcemia
c. Hypouricemia
d. Decreased urinary hydroxyproline level
e. Decreased uptake on radio-isotope bone scanning

15) You have recently attended a symposium about serum tumor markers. Which one of the following tumor markers has a fetal origin?

a. Prostate specific antigen
b. Calcitonin
c. CD34 antigen
d. Human chorionic gonadotrophin
e. CA-125

16) A 54-year-old man has been diagnosed with a malignant disease. He will receive chemotherapy. Which one of the following cytotoxic agents inhibits the enzyme topoisomerase II?

a. Daunorubicin
b. Irinotecan
c. Paclitaxel
d. 5-fluorouracil
e. Ifosfamide

17) A 28-year-old woman was brought to the Acute and Emergency Department. She has developed carbon monoxide poisoning. Which one of the following is *not* an indication for hyperbaric oxygen therapy in carbon monoxide poisoning?

a. Cerebellar dysfunction
b. Headache
c. Pregnancy
d. Carboxyhemoglobin level >40%
e. Seizures

18) A child of a Somalian family, was brought to the Emergency Department. You noticed protein-energy malnutrition. All of the following are suggestive of kwashiorkor rather than marasmus, *except?*

a. Bilateral pitting ankle edema
b. Hypoalbuminemia
c. Apathy
d. Abdominal distention
e. Severe muscle wasting

19) A 34-year-old man presented to the Emergency Room with a flare-up of his porphyria. All of the following are triggering factors for acute intermittent porphyria attacks, *except?*

a. Sodium valproate
b. Fasting
c. Propranolol
d. Sulfonamides
e. General anesthetics

20) A 32-year-old woman was referred to you because of repeated vomiting. You consider Bulimia nervosa. Which one of the following is *not* a feature of Bulimia nervosa?

a. Scratches on the dorsum of the hands
b. Upper gastrointestinal hemorrhage
c. Puffy cheeks
d. Amenorrhea in almost all cases
e. Teeth enamel erosions

21) A patient developed repeated chest infections. You ran a battery of investigation. His final diagnosis turned out to be an inherited immune deficiency syndrome. Which one of the following primary immunodeficiency disorders has an X-linked pattern of inheritance?

a. Hyper-IgM syndrome
b. Common variable immune deficiency
c. Reticular dysgenesis
d. DiGeorge syndrome
e. Chédiak-Higashi syndrome

22) You read a journal article about cytokines. Which one of the following cytokines is produced by macrophages?

a. IL-2
b. IL-5
c. Tumor necrosis factor beta
d. Interferon gamma
e. IL-1

23) A 34-year-old man has gastroesophageal reflux disease. Which one of the following medications does *not* reduce the lower esophageal sphincter pressure?

a. Progesterone

b. Theophylline
c. Propranolol
d. Cisapride
e. Verapamil

24) A 67-year-old man was brought to the Emergency room. He was diagnosed with congestive heart failure before 1 year. He has severe shortness of breath. The chest is full of crackles. Which one of the following factors does *not* result in cardiac decompensation in a previously well-controlled congestive heart failure?

a. Cardiac ischemia
b. Hyperthyroidism
c. Low dietary salt intake
d. Anemia
e. Fever

25) A 54-year-old man underwent chest CT scanning because of persistent unexplained cough. You skimmed the CT film and tried to remember the anatomy of the mediastinum. Which one of the following structures lies within the posterior mediastinum of the chest?
a. Thymus
b. Phrenic nerves
c. Hemiazygos vein
d. Innominate vein
e. Cricoid cartilage

26) A patient had developed recurrent nose bleeds and ecchymosis. Which one of the following is the *correct* statement with respect to Bernard-Soulier syndrome?

a. Has an X-linked recessive mode of inheritance
b. Microthrombocytes are seen on peripheral blood films
c. The bleeding time is marginally prolonged
d. The defect lies in the glycoprotein Ib/IX complex
e. Platelet transfusion is contraindicated

27) A 44-year-old man complains of excessive day-time somnolence. Which one of the following statements is *true* about obstructive sleep apnea?

a. Men are affected much less frequently than women
b. Obesity is noted in 90% of cases
c. Hypoxemia and hypercarbia occur in the minority of patients
d. Multiple sleep latency testing is the investigation of choice
e. The ESR is usually elevated

28) A 78-year-old woman fell on the ground and developed right-sided Colle's fracture. Which one of the following medications does *not* increase the bone mineral density:

a. Calcitonin
b. Raloxifene
c. Alendronate
d. L-thyroxin
e. Tibolone

29) Which one of the following statements about hand, foot, and mouth disease is the *correct* one?

a. Caused by human Herpes virus type 8
b. Adults are the usual victims
c. The oral lesions do not ulcerate
d. The disease is chronic and fluctuating
e. The hand rash is vesicular

30) A 45-year-old man presented with a localized area of scalp hair loss. You examined the patient and found scarring alopecia. Which of the following results in localized scarring alopecia?

a. Alopecia areata
b. Androgenetic alopecia
c. Discoid lupus erythematosus
d. Hyperthyroidism
e. Telogen effluvium

31) A 65-year-old man developed rapidly progressive dementia and myoclonus. You consider Creutzfeldt-Jacob disease (CJD). Which one of the following favors sporadic CJD over the new-variant CJD?

a. An earlier age of onset
b. Typical presentation with neuropsychiatric manifestations

c. The characteristic EEG changes
d. A slower pace
e. The non-fatal course

32) A 39-year-old woman presents with mooning of the face and central obesity. You consider Cushing's syndrome. Which one of the following is a screening test for Cushing's syndrome?

a. Random serum ACTH
b. Pituitary MRI scan
c. Adrenal vein sampling
d. Radio-cholesterol scanning
e. 24-hour urinary free cortisol

33) A 51-year-old man is about to undergo intravenous urography. Which one of the following is *not* a risk factor for contrast nephropathy?

a. Pre-existent renal impairment
b. Diabetes mellitus
c. Use of high osmolality contrast media
d. Multiple myeloma
e. Positive HBs antigen blood testing

34) A 19-year-old man was brought to the Emergency Department with several complaints. You did blood and urinary toxicology screen and found that he is a cocaine addict. Which one of the following is *not* a consequences of cocaine abuse?

a. Improvement of pre-existent asthma
b. Hyperthermia
c. Hyperprolactinemia
d. Pneumothorax
e. Intestinal ischemia

35) A 66-year-old man presented with easy forgetfulness. He scored 21 on mini-mental state examination (MMSE). Which one of the following is a potentially treatable cause of dementia?

a. Multi-infarct dementia
b. Huntington's disease
c. Lewy-body type of dementia

    d. Wilson's disease
    e. Idiopathic Parkinson's disease

36) A 40-year-old man developed recurrent hypoglycemia. All of the
    following can result in both fasting *and* reactive hypoglycemia, *except*?

    a. Insulinoma.
    b. Insulin auto-antibodies
    c. Insulin receptor auto-antibodies
    d. Adrenal failure
    e. Pentamidine

37) A 35-year-old female presented with pallor. Her serum ferritin is low.
    Which one of the following decreases serum ferritin level?

    a. Adult Still's disease
    b. Rheumatoid arthritis
    c. Chronic liver disease
    d. Hypothyroidism
    e. Frequent blood transfusion

38) A 36-year-old man was recently diagnosed with ulcerative colitis. All of
    the following parameters indicate the presence of severe attack of
    ulcerative colitis, *except*?

    a. Five bowel motions per day
    b. Frank blood in stool
    c. ESR of 50 mm/hour
    d. Serum albumin of 24 g/L
    e. Hemoglobin of 10.5 g/dl

39) A 29-year-old man presents with recurrent chest tightness, low-grade
    fever, headache, and malaise. Your final diagnosis is hypersensitivity
    pneumonitis (extrinsic allergic alveolitis). Which one of the following
    hypersensitivity pneumonitides is associated with exposure to
    Aspergillus species?

    a. Byssinosis
    b. Maple bark stripper's lung
    c. Malt worker's lung
    d. Bird fancier's lung

e. Humidifier's fever

40) With respect to the hemodynamic effects of inspiration, which one of the *correct* statement?

a. The JVP rises
b. The heart rate slows
c. The blood pressure rises
d. The second heart sound splits
e. The right ventricular preload deceases

41) Which one of the following conditions elevates serum ACTH levels?

a. Non-functioning pituitary macroadenoma
b. Hemochromatosis
c. 21-hydroxylase deficiency
d. Withdrawal of long-term prednisolone
e. Polycystic ovarian syndrome

42) A 33-year-old man presents with sore throat, pallor, and wide-spread ecchymosis. Your final diagnosis is aplastic anemia. All of the following can result in aplastic anemia, *except*?

a. Glue sniffing
b. Fanconi anemia
c. Paroxysmal nocturnal hemoglobinuria
d. Vitamin B$_{12}$ deficiency
e. Epstein-Barr viral infection

43) Which one of the following is the *correct* statement about cryoglobulins?

a. There are IgA or anti-IgE antibodies
b. Do not precipitate out in the cold
c. Type I cryoglobulinemia is associated with hepatitis C infection
d. Can result in peripheral neuropathy
e. Are monoclonal in type III cryoglobulinemia

44) Which one of the following does *not* result in skin Köebner's phenomenon?

a. Viral warts
b. Psoriasis
c. Lichen planus
d. Vitiligo
e. Basal cell papilloma

45) Which one of the following HLA molecules does *not* match the associated disease?

a. HLA B27-ankylosing spondylitis
b. HLA B5-Behçet disease
c. HLA DR4-rheumatoid arthritis
d. HLA DR2/DQ6-sporadic narcolepsy
e. HLA C3-hereditary hemochromatosis

46) Which one of the following statement is the *correct* one with respect to *P*-value?

a. Is not useful to medical researchers
b. A value of 0.05 is the cutoff for a significant value
c. Is calculated by dividing the mean over the median
d. Used only to accept the null hypothesis
e. Normally ranges from 0.00005 to 1.05

47) Systemic hypertension can safely be controlled during pregnancy using the all of the following medications, *except*?

a. Alpha methyldopa
b. Hydralazine
c. Ramipril
d. Labetalol
e. Nifedipine

48) Which one of the following is consistent with Staphylococcal scalded skin syndrome rather than toxic epidermal necrolysis?

a. Intra-epidermal line of cleavage
b. Oral mucosa involvement
c. Negative Nikolsky sign
d. Healing with scarring
e. High mortality rate

49) Which one of the following diseases can result in primary "AL" amyloidosis?

    a.   Chronic tuberculosis infection
    b.   Familial Mediterranean fever
    c.   Juvenile rheumatoid arthritis
    d.   Crohn's disease
    e.   Plasma cell dyscrasia

50) A 67-year-old man developed acute ischemic stroke. He is hypertensive and diabetic. Which one of the following is considered to be an exclusion criteria for the use of thrombolytic therapy in acute ischemic stroke?

    a.   No evidence of hemorrhage on doing brain CT scan
    b.   Patient's age equal to or greater than 18 years
    c.   Heparin use within 1 week
    d.   Seizures at stroke onset
    e.   Plasma glucose 210 mg/dl

51) A 67-year-old man visits the doctor's office because of fatigue and impotence. In spite of giving up smoking, his cough has worsened recently. The supine blood pressure is 140/90 mmHg, which drops down to 100/60 mmHg upon standing. Blood urea is 12 mg/dl and serum potassium is 3.0 mEq/L. The blood counts are normal. What is the likely diagnosis?

a.   Polyarteritis nodosa
b.   Nicotine replacement therapy side effects
c.   Lambert-Eaton myasthenic syndrome
d.   Squamous cell cancer of the lung
e.   Goodpasture syndrome

52) A 25-year-old college student presents with recurrent bouts of palpitations and syncope. Her resting 12-lead ECG shows marked left ventricular hypertrophy. The blood pressure is 120/75 mmHg. 24-hour Holter monitoring uncovers multiple attacks of ventricular tachycardia. What is the best intervention?

    a.   Wait and see
    b.   Daily oral amiodarone

    c.   Radiofrequency ablation of the aberrant pathway

    d.   Placement of a permanent pacemaker

    e.   Implantation of a cardioverter-defibrillator device

53) Because of developing headache and a short lapse of consciousness, a brain CT scan was done for a 54-year-old female. While skimming this CT scan film, you see a hyperdense lesion imparting a spider-legs pattern at the basal cisterns. What is the diagnosis?

    a.   Hemorrhagic choriocarcinoma secondary tumors

    b.   Acute subarachnoid hemorrhage

    c.   Chronic subdural hematoma

    d.   Acute epidural hematoma

    e.   This is a normal CT with extensive boney artifacts

54) While sitting in front of the TV watching a football game, a young man vomits a frank blood. You admit him to the Acute and emergency Department. He is conscious and afebrile. His blood pressure is 70/40 mmHg and the pulse rate is 130 beats/minute, which is regular. What is the best *next* step?

    a.   Doing upper GIT endoscopy with LASER therapy

    b.   Giving intravenous terlipressin

    c.   Arranging for balloon tamponade

    d.   Resuscitation

    e.   Giving oral propranolol

55) A middle-aged man with reasonably controlled chronic renal failure develops biochemical worsening. All of the following are potentially reversible causes of such worsening, *except?*

    a.   Hypertension

    b.   Hypotension

    c.   Tetracycline therapy

    d.   Fever

    e.   Hypoglycemia

56) A 24-year-old homosexual man is suspected to have acute HIV seroconversion illness after returning from a trip to Asia. Which one of the following is a well-documented clinical feature?

a. Oral ulcers
b. Chronic hepatic failure
c. Acute renal shutdown
d. Dactylitis
e. Duodenal fistula formation

57) A 21-year-old vagrant male developed methanol intoxication. Which one of the following is *not* used as a mode of treatment in this patient?

a. Oral ethanol
b. Intravenous ethanol
c. Hemodialysis
d. Oral ethylene glycol
e. Intravenous 4-methylpyrazole

58) A blood film that belongs to a child with acute favism is likely to show which one of the following?

a. Burr cells
b. Acanthocytes
c. Bite cells
d. Basophilic stippling
e. Smudge cells

59) A teenage girl has developed recurrent meningococcal meningitis. No anatomical defect was detected in the head or spine. You suspect some sort of immune deficiency. Which one of the following complement component deficiencies results in this type of recurrent infection?

a. C1q esterase inhibitor deficiency
b. C6 deficiency
c. C2 deficiency
d. C1s deficiency
e. C10 deficiency

60) A 21-year-old woman visits the physician's office for a medical advice. She has idiopathic grand mal epilepsy for which she takes sodium valproate. She is sexually active and takes oral contraceptive pills. She does not want to get pregnant. You plan to change her current antiepileptic medication to which one of the following?

a. Carbamazepine
b. Phenytoin
c. Phenobarbitone
d. Lamotrigine
e. Primidone

61) A 21-year-old college student female presents with recurrent syncope over the past 2 years. Her roommate says that this has occurred hundreds of times. There is no family history of a similar condition and she denies drug abuse. Her clinical examination is unremarkable. What do you think she has?

a. Mitral valve prolapse
b. Generalized catamenial epilepsy
c. Non-epileptic attacks
d. Primary pulmonary hypertension
e. Carotid sinus hypersensitivity

62) A 54-year-old female with long standing rheumatoid arthritis presents with bilateral leg swelling that pits on pressure. Her cardiovascular system examination is unremarkable. Choose an investigation to be done next?

a. Echocardiography
b. Liver function testing
c. Subcutaneous fat aspiration
d. 24-hour urinary protein
e. Rheumatoid factor titer

63) A 68-year-old man presents with wide-spread skin lesions in the form of red raw erosions and flaccid bullae. He denies chronic diseases or being on medications. Where would you like to examine further?

a. Perinatal area
b. Scalp
c. Natal cleft
d. Posterior neck
e. Mouth

64) A 69-year-old woman presents with generalized seizures. She has chronic obstructive airway disease for which she takes many medications, including inhalers. One week ago, she took oral ciprofloxacin to treat her urinary tract infection. What is the cause of this new seizure?

    a. Hypoxic encephalopathy
    b. Hemispheric brain tumor
    c. Theophylline toxicity
    d. Ciprofloxacin intoxication
    e. Hyponatremia

65) A middle-aged man presents with headache for 1 year. Brain MRI reveals a pituitary macroadenoma. His serum prolactin is 2300 mU/l. What is the cause of this abnormal blood test?

    a. Stress
    b. Pituitary prolactinoma
    c. Disconnection hyperprolactinemia
    d. Secondary brain tumor
    e. Metoclopramide therapy

66) A 69-year-old female has a blood pressure of 180/110 mmHg. Her blood pressure has been always high over the past 3 months. Her general practitioner is unable to lower it within the recommended target range. Her hypertension was diagnosed 10 years ago. The right kidney is found to be smaller than the left one upon doing abdominal ultrasonography. The pedal pulses are feeble bilaterally. What is the reason for this high blood pressure?

    a. Reflux nephropathy
    b. Fibromuscular dysplasia-associated renal artery stenosis
    c. Atherosclerotic renal artery stenosis
    d. Adult polycystic kidneys disease
    e. Congenital renal hypoplasia

67) A 70-year-old man has developed acute right-sided painful knee. Joint aspiration confirms the presence of negatively birefringent crystals. He has many chronic illnesses, for which he takes multiple daily medications. Which medication is responsible for this presentation?

a. Aspirin, 1000 mg/day
b. Tolbutamide
c. Gabapentin
d. Aspirin, 81 mg/day
e. Atenolol

68) A 10- year-old child presents with fever, headache, vomiting, and neck stiffness. A provisional diagnosis of meningitis is made, and you plan to do lumber puncture. Your spinal needle will pass through all of the following structures, *except*?

a. Supraspinous ligament
b. Infraspinous ligament
c. Dura
d. Anterior longitudinal ligament
e. Ligamentum flavum

69) Because of a 2-week history of voice hoarseness, a 14-year-old girl is referred to you for further evaluation. She has a small thyroid mass, and you consider thyroid malignancy. Which type of malignancy most likely she has?

a. Anaplastic carcinoma
b. Papillary carcinoma
c. Squamous carcinoma
d. Clear cell carcinoma
e. Follicular carcinoma

70) A 54-year-old salesperson was fired from her job because of repeated absences. She states that she feels tired all the time, and this has not improved after taking a nap or a long sleep. Her routine bloods, serum TSH, and polysomnography are normal. You consider chronic fatigue syndrome. How would you advise her?

a. Spend more time in bed
b. Undertake graded exercises
c. Eat food rich in unsaturated fatty acids
d. Start relaxation techniques
e. Enroll in a support group

71) A 43-year-old male develops acute atrial fibrillation. He asks if there is any medical treatment that can reverse his fast heart rate, because he is afraid from the DC shock. You tell him that there are some cardiac medications that can terminate atrial fibrillation. What would you give for this purpose?

    a.   Metoprolol
    b.   Verapamil
    c.   Nifedipine
    d.   Sotalol
    e.   Flecainide

72) A 12-year-old child develops acute renal failure, 1 week after a short-lived bloody diarrhea. He had no severe dehydration at that time. Schistocytes are seen on peripheral blood film. Which one of the following is *consistent* with your provisional diagnosis?

    a.   Polychromasia
    b.   Prolonged PT
    c.   Thrombocytosis
    d.   Prolonged aPTT
    e.   Normal hemoglobin value

73) A 34-year-old man has bipolar disorder and he takes daily lithium. Today, he was admitted to the Acute and Emergency department with nausea, vomiting, coarse tremor, and confusion. His brother says, "His general practitioner added a new medication last week for certain reason." What do you think the name of this new medication is?

    a.   Sodium valproate
    b.   Paracetamol
    c.   Hydrochlorothiazide
    d.   Carbamazepine
    e.   Fluoxetine

74) A 42-year-old man presents with watery diarrhea that is not responding to anti-motility agents. He underwent extensive small bowel resection following an embolic mesenteric infarction 3 months ago. Stool examination does not reveal any pathogen. How would you treat?

a. Metronidazole
b. Cholestyramine
c. Warfarin
d. Vitamin B_{12} injections
e. Tetracycline

75) A 33-year-old woman, who has vague personality changes, presents with gross hematuria. Further work-up reveals renal vein thrombosis. Examination reveals livedo reticularis. Which of the following serum antibodies is likely to be positive in this woman?

a. Anti-Jo1 antibody
b. Anti-Sm antibody
c. Anti-phospholipid antibodies
d. Anti-ribonucleoprotein antibodies
e. Anti-SSA antibodies

76) A 39-year-old male presents with chronic cough and exertional shortness of breath. Chest examination is consistent with hyperinflation. Chest X-ray is suggestive of predominantly lower lobes emphysema. You suspect alpha-1 antitrypsin deficiency. His alpha-1 antitrypsin phenotype is?

a. SS
b. ZM
c. SM
d. ZZ
e. MM

77) A 61-year-old woman visits the physician's office for her annual check-up. She denies symptoms and she takes no medications. She neither smokes nor drinks alcohol. The clinical examination is unremarkable. Her blood tests are normal, apart from slightly raised serum calcium. What is the likely cause of this hypercalcemia?

a. Primary hyperparathyroidism
b. Secondary hyperparathyroidism
c. PTH-related peptide secretion by an occult cancer
d. Surreptitious vitamin D ingestion
e. Sarcoidosis

78) A 32-year-old man has long-standing type I diabetes mellitus. He has demonstrated 3 positive urine tests for microalbuminuria over a 1 year period of follow-up. His blood pressure is 120/76 mmHg. What would you do?

   a. Continue the follow-up
   b. Give enalapril
   c. Do renal biopsy
   d. Refer for renal transplantation
   e. Intensify his blood glucose control

79) A 71-year-old man consults a local ophthalmologist because of poor vision. Which one of the following ocular findings is *not* a normal age-related change?

   a. Bilateral ptosis
   b. Small pupils
   c. Poor accommodation
   d. Impaired up-gaze
   e. Down-beating nystagmus

80) A 65-year-old man complains of recurrent lightheadedness and dimming of vision. His clinical examination is unremarkable. 24-hour Holter monitoring suggests sick sinus syndrome as the cause of these episodes. What is the best treatment?

   a. Wait and see
   b. Implantation of a permanent pacemaker
   c. Radiofrequency ablation
   d. Cardioverter-defibrillator placement
   e. Fludrocortisone

81) A 28-year-old woman has had her hair dyed 2 days ago. Today, she presents with an itchy papulovesicular rash on the upper face and scalp. What does the woman have developed?

   a. Seborrheic dermatitis
   b. Atopic dermatitis
   c. Contact dermatitis
   d. Tinea capitis
   e. Discoid lupus

82) A 16-year-old high school student is brought by his parents to consult you. The boy has pressure of speech and grandiosity, and is easily irritated. The mother says her son had periods of low mood, sadness, insomnia, and death ideas. What does the boy suffer from?

   a. Chronic schizophrenia
   b. Major depression
   c. Acute mania
   d. Substance abuse
   e. Bipolar disorder

83) A 46-year-old man, who has been diagnosed with chronic obstructive airway disease, presents to the emergency room with increased shortness of breath and fever for 3 days. You are thinking of acute exacerbation of his chest problem due to chest infection. In spite of optimal medical treatment, he deteriorates rapidly. His blood pH is 7.15. What is the most appropriate step for the time being?

   a. Increase the dose of steroids
   b. Add intravenous magnesium
   c. Invasive ventilation
   d. Giving high flow, high concentration O_2
   e. Intravenous doxapram

84) Few days after normal vaginal delivery of a healthy-looking baby, a 20-year-old woman develops drowsiness and paraparesis. She is afebrile. Fundoscopy reveals florid papilledema. No neck stiffness is detected. What is the cause of this deterioration?

   a. Pyogenic meningitis
   b. Superior sagittal sinus thrombosis
   c. Herpes simplex encephalitis
   d. Puerperal sepsis
   e. Watershed infarctions

85) After recovering from extensive anterior wall myocardial infarction, a 63-year-old man develops recurrent presyncope. His 12-lead ECG reveals complete heart block in addition to the recent infarction changes. What would expect to find on auscultating his precordium because of this heart block?

a. Reversed splitting of the second heart sound
b. Mitral regurgitation
c. Tricuspid regurgitation
d. Variable intensity of the first heart sound
e. Forth heart sound

86) You conduct a research about a novel medication acting on blood pressure, and you choose a double-blind, placebo-controlled method of trial. What is the benefit of this double blinding?

   a. To eliminate the subjective bias
   b. To increase the number of participations
   c. To double the number of researchers
   d. To decrease the expenses
   e. To boost the propaganda

87) A 26-year-old woman comes to the emergency room dyspnic and tachypnic. Pulse oximetry reveals a saturation of 88%. Chest examination is unremarkable, as is her plain chest X-ray film. One week ago, she gave birth to a 3 Kg full-term baby and there were no complications at that time. What will you do next?

   a. Repeat the chest X-ray in expiration
   b. Start enoxaparin
   c. Arrange for V/Q lung scanning
   d. Send for spiral CT chest scanning
   e. Check plasma D-dimer level

88) A 73-year-old man complains of retrosternal chest pain and odynophagia for 4 weeks. He sustained an osteoporosis-related hip fracture 4 months ago, for which he takes a daily medication. What medication is responsible for his new presentation?

a. Calcitonin
b. Calcium
c. Vitamin D
d. Alendronate
e. Testosterone skin patch

89) A 32-year-old businessman presents with non-itchy generalized maculopapular skin rash for few days. The palms and soles are involved and you find generalized lymph node enlargement. His liver enzymes are raised. What test you would order?

    a.   Antinuclear antibody
    b.   Serum VDRL
    c.   Serum ELISA for HIV
    d.   Skin biopsy
    e.   Serum hepatitis B virus surface antigen

90) A 21-year-old male has recurrent renal calculi because of idiopathic hypercalciuria. Which one of the following medications would be beneficial in preventing further stone formation?

    a.   Frusemide
    b.   Calcitonin
    c.   Aluminum hydroxide
    d.   Hydrochlorothiazide
    e.   Risedronate

91) A 21-year-old vagrant male is brought to the Acute and Emergency department with fever, rigor, and shortness of breath. Examination reveals signs of severe tricuspid regurgitation. The ESR is high and C-reactive protein is 110 mg/dl. What is the diagnosis?

    a.   Brucella endocarditis
    b.   Staphylococcal endocarditis
    c.   Chrodae rupture of tricuspid valve prolapse
    d.   Embolus in the right ventricular outflow tract
    e.   Arterialization of the upper right ventricle

92) A 49-year-old man has been diagnosed primary Sjögren's syndrome 2 years ago because of dry eyes and mouth as well as parotid enlargement. Today, she says that she has been experiencing intermittent renal colic, every now and then. The new presentation is due to?

    a.   Alimentary hyperoxaluria
    b.   Recurrent urinary tract infections
    c.   Distal renal tubular acidosis

    d.   Drug-induced side effects

    e.   Conversion disorder

93) A 12-year-old girl has a sever attack of asthma that is not responding to conventional medical treatment. What is the best *next* step?

    a.   Increase the concentration of inspired O_2

    b.   Add intravenous magnesium

    c.   Endotracheal intubation and mechanical ventilation

    d.   Give intravenous aminophylline

    e.   Shift oral prednisolone to intravenous hydrocortisone

94) A 21-year-old man has recurrent lancinating pains in the area of right maxilla and mandible. These pains are brought about by tooth brushing and laughing. Which one of the following is *not* consistent with the diagnosis of idiopathic trigeminal neuralgia?

    a.   The patient's age

    b.   The localization of the pain

    c.   Triggering by tooth brushing

    d.   The pain type

    e.   The pain frequency

95) A 67-year-old male was recently been diagnosed with small-cell lung cancer of the upper part of the left lung. The surgeon is thinking about doing surgery for the tumor and is arranging for further tests of operability. Which one of the following factors would prevent the surgeon form doing the surgery?

    a.   Involvement of the aorta

    b.   Involvement of the vertebral column

    c.   The tumor site

    d.   The tumor type

    e.   The patient's age

96) A 51-year-old woman had chronic sinusitis and asthma. She developed a palpable pupuric rash as well as median and tibial nerve palsies. These have developed after stopping her oral prednisolone tablets, which were prescribed for her asthma. Her blood eosinophil count is $2000/ml^3$. What does the woman have?

a. Allergic bronchopulmonary aspergillosis
b. Chronic eosinophilic pneumonia
c. Visceral larva migrans
d. Churg-Strauss vasculitis
e. Microscopic polyangiitis

97) An 11-month-old infant develops weight loss, failure to thrive, and bulky stool. His Serum IgA anti-gliadin and anti-endomysial antibodies are negative. Duodenal biopsy reveals subtotal villous atrophy. He makes an excellent degree of improvement after starting gluten-free diet. What is the cause of the negative antibody testing?

a. He does not have celiac disease
b. He was in remission
c. Co-existing selective IgA deficiency
d. Low titer of the tested antibodies
e. The presence of blocking antibodies

98) A56-year-old patient is referred form the psychiatry department to you because of jaundice and itching. He takes multiple daily medications. His viral hepatitis screen is negative. The patient denies abdominal pain or nausea. What is the best step to be undertaken for the time being?

a. Repeat the viral hepatitis serum markers after 2 to 4 weeks
b. Stop phenothiazines
c. Start D-penicillamine
d. Perform Coomb's test
e. Wait and see

99) A 42-year-old uremic patient is being evaluated for allogenic kidney transplantation. Which one of the following rejection reactions is caused by preformed antibodies?

a. Chronic
b. Acute
c. Hyperacute
d. Accelerated
e. Delayed chronic

100)   A 15-year-old male visits the doctor's office with his father for an advice. The father says that boy's feet are high-arched and their sole is distinctly hollow when bearing weight. You find bilateral pes cavus and thin stork-like legs. What does the boy have?

a.   Fabry disease
b.   Charcot-Marie-Tooth disease
c.   Tabes dorsalis
d.   Gaucher's disease
e.   Congenital hip dislocation

This page was intentionally left blank

Mock Paper Number One

Answers

This page was intentionally left blank

1) c.

Objective: review neurological localization of various signs and symptoms in neurology.

Aphasia is a supra-tentorial, dominant hemispheric, and cortical sign; it rarely occurs with subcortical lesions, e.g., so-called thalamic aphasia. Apraxia is a supra-tentorial hemispheric sign; there are many types of apraxia, each points out to a different localization. Damage to both "leg" cortical areas, e.g., in midsagittal frontal meningioma, can result in paraparesis. Paraparesis also occurs with spinal cord lesions at the level of the dorsal spine (T_{1-12}). Brown-Séquard syndrome is a hemi-sectioning of the spinal cord; the brainstem is not involved. Damage to the basis pontis, e.g., by tumor or hemorrhage, may result in locked-in syndrome (a de-efferentation state, which superficially resembles coma).

2) d.

Objective: risk stratification of unstable angina and the corresponding management option(s).

Risk stratification in unstable angina is important as part of the overall management. This will guide you further, whether to continue on medical treatment only (for low risk group) or to go for a more invasive and interventional step (cardiac catheterization for high risk group).

3) a.

Objective: usefulness of radioiodine uptake test in the differential diagnosis of hyperthyroidism.

Graves' disease, toxic multi-nodular goiter, and solitary toxic adenoma (which are the commonest causes of hyperthyroidism) result in increased uptake on thyroid isotope scanning; this also applies to TSH-induced thyrotoxicosis and follicular thyroid cancer. The other causes of hyperthyroidism (such as factitious hyperthyroidism, struma ovarii,…etc.), which are rare in clinical practice, are responsible for hyperthyroidism with negligible radioiodine uptake.

4) e.

Objective: review risk factors for the development of esophageal cancers.

The other risk factors for esophageal squamous cell cancer are alcohol excess, presence of post-cricoid web, cigarette smoking, and chewing tobacco or betel nuts. Esophageal adenocarcinoma has an association with gastroesophageal reflux disease.

5) a.

Objective: review medical therapy of different phases of acute leukemia.

Many medications can be used in different phases of leukemia treatment. Some medications are given intravenously and some are administered orally, while others are used intrathecally. Vincristine should never be given intrathecally.

6) a.

Objective: review the clinical features of Wegener's granulomatosis and its mode of presentation.

Ocular inflammation (conjunctivitis, episcleritis, scleritis, uveitis, and exophthalmos), skin rashes, sinusitis, pulmonary infiltrates (on plain chest X-ray), cough, hemoptysis, epistaxis, shortness of breath, chest discomfort, renal failure, arthritis or arthralgia, and pleuritis all may be encountered. Fever, weight loss, anorexia, and malaise are non-specific features of vasculitides.

7) a.

Objective: review the causes of unexplained fever in HIV-infected patients.

Unexplained fever is a fever that has no localizing signs. Disseminated Mycobacterium avium complex (MAC) infections can be another cause; the presence of isolated elevation of serum alkaline phosphatase in any HIV-infected patient with a CD4 positive count of $< 50/mm^3$ should always prompt a search for MAC infection. Fever alone may the presenting feature of acute HIV retroviral seroconversion illness.

Aggressive high-grade non-Hodgkin's lymphoma should not be overlooked. Oral candidiasis, Cryptosporidium diarrhea, and HIV nephropathy do not result in fever.

8) a.

Objective: review the mode of action of anti-arrhythmic medications and their classification.

Flecainide and propafenone are class Ic agents while procainamide, quinine, and disopyramide belong to class Ia. Class Ib encompasses tocainide, phenytoin, lidocaine, and mexiletine. Together with amiodarone and bretyllium, sotalol is a class III agent; this class prolongs the plateau phase of the action potential. Class IV drugs are the calcium channel blockers; verapamil and diltiazem. Beta blockers belong to class II. Digoxin does not follow this classification of Vaughn-Williams.

9) c.

Objective: review the general classification (pre-renal, renal, and post-renal) and causes of acute renal failure.

Iodinated contrast agents may result in nephrotoxic acute tubular necrosis; gentamicin, amphotericin B, mercury, and carbon tetrachloride may also cause this renal damage. Bladder neck obstruction is a post-renal cause of azotemia. Protracted vomiting and any cause of low circulating blood volume (such as severe diarrhea, hemorrhage, and aggressive diuresis) may end-up with pre-renal failure. Acute renal arterial occlusion (e.g., bilateral thromboembolism) can result in pre-renal failure. Malignant hypertension is an intrinsic cause of renal impairment. Myoglobinuria, hemoglobinuria, and myeloma light chains are endogenous nephrotoxins which can damage the renal tubules.

10) c.

Objective: review the cause of elevated serum amylase in various disease states.

The serum levels of amylase of pancreatic origin may be elevated after opiate administration, gastric and small bowel perforation, mesenteric infarction, and post-ERCP.

Serum amylase of non-pancreatic origin may be elevated in lactic acidosis, salivary gland adenitis, diabetic ketoacidosis, renal failure, "ruptured" ectopic pregnancy, and post-operative states.

11) b.

Objective: review the protean complications of acute hepatic failure and their treatment modalities.

The filed liver would be unable to generate gluconeogenesis and cannot destroy insulin; therefore, hypoglycemia may ensue rapidly. Coagulopathy, DIC, and GIT hemorrhage (gastric erosions are common) all may complicate liver failure; thrombocytosis is not a feature. The hallmark of acute hepatic failure is encephalopathy. Confusion and agitation are very common and may be due to hypoglycemia, hypoxia, cerebral edema, hepatic encephalopathy, or CNS hemorrhage.

12) b.

Objective: review the causes of hypophosphatemia and its consequences in the short and long-term.

In addition, prolonged or severe hypophosphatemia may cause muscle pain, weakness, and raised serum CPK. Red cell hemolysis as well as white cell and platelet dysfunction may occur. Rickets may develop in the long-term in children. Cardiac dysfunction and arrhythmia are well-recognized features. The resulting encephalopathy can cause seizures, coma, agitation…etc.

13) b.

Objective: review the characteristic features of cluster headache and how to differentiate it from migraine (with or without aura).

Cluster headache characteristics:

1- Severe unilateral orbital, supraorbital and/or temporal pain that lasts from 15 to 180 minutes, often at the same time each day.
2- The headache should be associated with at least one of the following "signs" on the side of the pain: ptosis, pupillary constriction, conjunctival

injection, lacrimation, nasal congestion, rhinorrhea, eyelid edema, and forehead (and facial) sweating.

3- Frequency: from one attack every other day to eight attacks per day.

4- Should occur in clusters of 3-6 weeks. Rarely chronic.

14) a.

Objective: review the investigations for Paget's disease of the bone and their characteristic results.

Raised serum level of the bone fraction (isozyme) of alkaline phosphatase may be the only clue; together with serum osteocalcin, are markers of disease activity. Hypercalcemia and hypercalciuria develop in immobilized patients; these call for medical treatment. Raised serum uric acid reflects the increased cell turnover; it is rarely clinically significant. There is elevation in urinary hydroxyproline level; a marker of increased bone resorption. Bone scanning is the most efficient way for surveying Pagetic sites; the isotope uptake is increased in active areas.

15) e.

Objective: review the "indications" and "usefulness" of tumor markers in various malignancies.

Serum prostate specific antigen is elevated in prostatic cancer and benign hyperplasia; it may also be elevated in certain breast cancers. CA-125 is elevated in epithelial ovarian cancers; its place in the screening of high risk females is still controversial. Alpha fetoprotein levels become prominent in germ cell tumors and primary hepatocellular carcinoma. Elevated serum level of human chorionic gonadotrophin (hCG) is encountered in choriocarcinoma, teratoma, and certain gastrointestinal (GIT) tumors. Carcinoembryonic antigen (CEA) serum level is raised in malignancies of the GIT, breast, and lung. Note that the above tumor markers (apart from hCG) may be detected in blood samples of healthy individuals, but at a very low level. Alpha fetoprotein, CEA, and CA-125 are actually fetal antigens. Calcitonin is a marker of medullary thyroid carcinoma. CD34 antigen is a hematopoietic progenitor cell antigen; useful in hemangiopericytoma/solitary fibrous tumor, gastrointestinal stromal tumor, and dermatofibrosarcoma protuberance.

16) a.

Objective: review the classification and mode of action of various cytotoxic agents.

Etoposide, epirubicin, and daunorubicin inhibit the enzyme topoisomerase II. Irinotecan and topotecan block topoisomerase I. The taxanes (docetaxel and paclitaxel) are inhibitors of mitotic spindle assembly. 5-FU (5-flourouracil) is an anti-metabolite; the same group of methotrexate, cytarabine, 6-MP, and thioguanine. Cyclophosphamide, ifosfamide, and melphalan are alkylating agents.

17) b.

Objective: review the various methods used for improving oxygenation in acute CO poisoning and their rapidity of lowering carboxyhemoglobin level.

The earliest features of acute CO poisoning are headache, nausea, vomiting, unsteadiness, and nystagmus; these usually resolve somewhat quickly when the patient is moved away from the source of exposure. Although the use of hyperbaric O_2 therapy is still controversial, it is used in the presence of focal neurological signs (especially cerebellar signs), coma, carboxyhemoglobin blood level >40%, and pregnancy (it may result in abortion or premature labor because of fetal hypoxia).

18) e.

Objective: differentiate between kwashiorkor and marasmus.

The abdomen is distended mainly because of weakness and flaccidity of the anterior abdominal wall muscles. In marasmus, the abdomen may show some degree of distension; however, it is usually scaphoid. The kwashiorkor diet has low protein to energy ratio resulting in an increase in serum insulin and a fall in serum cortisol. These impair the hepatic capacity to synthesize albumin. In addition, many amino acids are being diverted to form acute phase reactants at times of infection. The kwashiorkor child appears miserable; the marasmic one is irritable, feels hungry, and constantly asks for food. Mild or no skin and hair changes occur in marasmus. The kwashiorkor hair is thin, sparse, and easily plucked, which may be grey or even reddish in color, and its skin can show hyperpigmentation, cracks, and denuded areas. The marasmic child lives on his own meet; severe wasting is the rule.

In kwashiorkor, some degree of wasting can be seen around the shoulders and upper arms, but severe wasting is highly unusual. However, many physical signs can be seen in both of these protein-energy malnutrition cases; this is especially true in severe and advanced cases.

19) c.

Objective: review the precipitating factors for acute intermittent porphyria (AIP) attacks and safe medications which are used in alleviating these attacks or treat co-morbidities.

Gabapentin can safely be used to control seizures; phenytoin and phenobarbitone may precipitate acute attacks, however. The other "unsafe" medications are sulfonamides, oral contraceptive pills, progestins, griseofulvin, tolbutamide, chlorpropamide, danazol, dapson, and sodium valproate. Fasting and diets low in carbohydrates may flare-up the disease; at least, 300 g/day of dietary carbohydrates should be given during attacks. Beta blockers are safe agents to control the hyper-adrenergic manifestations (hypertension and tachycardia). Narcotic analgesics are effective and safe in alleviating pain symptoms during acute attacks. General or local anesthesia may trigger an attack.

20) d.

Objective: review various signs of bulimia nervosa, and differentiate this condition from anorexia nervosa.

Scratches or calluses can be seen on the dorsum of the hands because of abrasion from teeth during induced vomiting. Frequent induced gagging may produce gastric or esophageal perforations or tears, hemorrhage, pneumomediastinum, or subcutaneous emphysema. The cheeks look puffy in half of patients because of bilateral parotid enlargement; serum amylase may be raised as well. Although menstrual irregularities are very common, frank amenorrhea is very rare (in contrast to anorexia nervosa where amenorrhea is one of the core features). The teeth enamel is eroded because of acid-induced dissolution and decalcification. Because bulimics are not wasted, physical findings may be subtle or even absent.

21) a.

Objective: review the classification, mode of inheritance, basic genetic and functional defect, and characteristic clinical features of primary immunodeficiency diseases.

The following primary immunodeficiency disorders have an X-linked inheritance pattern: X-linked (Bruton's) agammaglobulinemia, X-linked hyper-IgM syndrome, X-linked lymphoproliferative syndrome, a subset of severe combined immune deficiency syndrome, a subset of chronic granulomatous disease of childhood, Wiscott-Aldrich syndrome, and properdin deficiency. Although many cases of common variable immune deficiency and selective IgA deficiency are familial, the pattern of inheritance is unknown. DiGeorge syndrome is a sporadic "chromosomal microdeletion" syndrome. Severe combined immune deficiency has many subsets, with various genetic and functional defects. Reticular dysgenesis is an autosomal recessive disorder. Most of these diseases are autosomal recessive involving cell-mediated immunity, antibody production, and phagocytic function.

22) e.

Objective: review cytokines, their source, and their biological activities.

IL-2 is produced by T-cells and stimulates the proliferation and differentiation of T-cells, activates cytotoxic T-cells and NK cells, and co-stimulates B-cells and antibody production. IL-5 is formed by T-cells and stimulates the proliferation of eosinophils. Tumor necrosis factor beta is produced by T-cells (its old name is lymphotoxin); it activates endothelial cells, granulocytes, and B-cells while it inhibits angiogenesis and it is cytotoxic to many cell types. Interferon gamma is produced by T-cells and NK cells; it activates NK cells, T-cells, endothelial cells, and macrophages and has an anti-tumor activity, inhibits T-cell proliferation, and co-stimulates B-cell proliferation. IL-7 is produced by thymic strand cells; it is involved in the proliferation and differentiation of pre-B cells as well as proliferation of B-cells. The following cytokines are produced by macrophages: IL-1 (alpha and beta), IL-3, IL-6, IL-8, and TNF alpha (cachectin).

23) d.

Objective: review various causes of gastroesophageal reflux disease and factors that interfere with its treatment.

Cimetidine and omeprazole have nothing to do with the lower esophageal sphincter (LES) pressure.

Cisapride is a promotility agent that increases the LES pressure and augments gastric emptying; however, it is not used any longer because of its adverse cardiac effects.

24) c.

Objective: review the factors which may aggravate heart failure and interfere with its treatment.

The other factors are infection (especially chest infections), cardiac ischemia, hypertension, poor compliance with medications, Paget's disease of the bone, pregnancy, renal failure, pulmonary emboli, fluid overload, and high dietary salt intake.

25) c.

Objective: review the anatomical compartments of the chest and localization of tumors that arise within these compartments.

The esophagus and thoracic duct lie in the posterior mediastinum. The vagus nerves pass, at first, through the middle mediastinum to enter the posterior one, while the phrenic nerves are located in the middle mediastinum. The azygos and hemiazygos veins as well as sympathetic chains are within the posterior mediastinum. The thymus and the aortic arch are located in superior/anterior mediastinum. The cricoid cartilage is in the neck!

26) d.

Objective: review the inherited disorders of platelet function.

Bernard-Soulier syndrome is an autosomal recessive disorder. Giant platelets or macrothrombocytes can be seen on peripheral blood films. The bleeding time is markedly prolonged because of failure of platelets to adhere properly to the subendothelium. The defect lies in the glycoprotein Ib/IX complex; gp IIb/IIIa complex is abnormal in Glanzmann's. thrombosthenia. The hemorrhagic manifestations respond favorably to platelets transfusion; since alloantibodies can develop to the absent glycoprotein complexes in these patients, platelets transfusion should be considered only in urgent clinical situations.

27) c.

Objective: review sleep apnea disorders, their diagnostic testing and criteria, and treatment modalities.

Obstructive sleep apnea usually targets man. It affects about 4% of the general population. Obesity is noted in 50% of cases only; although obesity is a major risk factor, not all patients are obese or even overweight, and the mechanism by which obesity predisposes to such a disease is still unknown. Advanced cases may show pulmonary hypertension and right-sided heart failure as well as systemic hypertension. Many patients show inability to concentrate and may display depression, irritability, and even personality changes. Multiple sleep latency testing is used in suspected cases of narcolepsy. Full-night polysomnography is used to document the presence of sleep apnea. Clinical examination is usually unremarkable; however, there may be obesity as well as crowded and hyperemic posterior pharynx.

28) d.

Objective: review the medical treatment of osteoporosis.

All anti-osteoporosis medications increase the bone mineral density (BMD). Bisphosphonates, raloxifene, calcitonin, hormonal replacement therapy (HRT), and calcium (alone, without vitamin D) reduce the risk of vertebral fractures (usually of the dorsal spine). Bisphosphonates, HRT, and calcium plus vitamin D reduce the risk of non-vertebral fractures (of the hips and wrists).

29) e.

Objective: review hand, foot, and mouth disease

This disease is caused by Coxackie A16 viral infection. Children are the main target of the virus, therefore, resulting local or household outbreaks. The oral lesions ulcerate rapidly after their appearance. The disease is self-limiting, usually within 1-2 weeks. The skin rash is vesicular; the lesions are usually painful and may require analgesia.

30) c.

Objective: review the causes of localized versus generalized alopecia, and whether they are scarring or not.

Androgenetic alopecia, telogen effluvium, alopecia areata, and hyperthyroidism are non-scarring causes of scalp hair loss. Localized scarring alopecia may be caused by discoid lupus, Herpes zoster, pseudopelade, kerion, and certain developmental defects. Generalized scarring alopecia may result from discoid lupus, lichen planus pillaris, folliculitis decalvans, and scalp radiotherapy.

31) c.

Objective: compare the sporadic CJD with its new-variant.

The sporadic disease usually targets people during their 7th decade; the characteristic feature of the new-variant form is the development in younger people. The new-variant typically presents with neuropsychiatric manifestations rather than with rapidly progressive dementia and startle myoclonus. The new-variant has no characteristic EEG features; the typical EEG changes of the sporadic CJD are seen only in 65% of cases and are usually transient. The sporadic CJD patient usually dies within 6-8 months of diagnosis; this figure rises to one year in the new-variant form. Both forms are fatal.

32) e.

Objective: review the diagnostic algorithm for non-iatrogenic Cushing's syndrome.

In general, all of the investigations mentioned in the question are used in the diagnostic plan, but only the last one is used for "screening". Generally, 4 tests are applied to screen for Cushing's syndrome, and these are: overnight low-dose dexamethasone suppression test; 24-hour urinary free cortisol; 48-hour low-dose dexamethasone suppression test; and circadian plasma cortisol (midnight and morning samples).

33) e.

Objective: review the risk factors for contrast nephropathy and how to prevent it.

The well-recognized risk factors for contrast nephropathy are pre-existing renal impairment, diabetes mellitus (especially those take metformin), use of high osmolal media, and myeloma patients.

34) a.

Objective: review the complications of cocaine abuse.

The list is very long. Cocaine abuse is one of the commonest causes of drug-related visits to the emergency rooms. It exacerbates asthmatic symptoms and produces hyperthermia. It can result in sexual dysfunction unrelated to hyperprolactinemia. Besides pneumothorax, it may lead to pneumomediastinum, pneumoperitonium, pulmonary edema, pulmonary hemorrhage, bronchiolitis obliterans, and "Crack lung". It may also lead to gastrointestinal perforation and colitis.

35) d.

Objective: review the causes of potentially treatable versus non-treatable dementias.

Around 10% of all dementia are potentially treatable or somewhat reversible. This list includes vitamin B_{12} deficiency, normal pressure (communicating) hydrocephalus, intracranial mass lesion, Wilson's disease, tertiary syphilis (general paresis of insane), chronic hepatic encephalopathy, granulomatous meningitis, and chronic medications (drugs or alcohol abuse). The commonest causes of dementia (Alzheimer's disease, Lewy's dementia, and multi-infarct demnetia) are irreversible.

36) e.

Objective: differentiate between "fasting" and "reactive" hypoglycemia and review the conditions that can result in both.

Insulinoma, insulin auto-antibodies, insulin receptor auto-antibodies, and adrenal failure are the causes of *both*, fasting and reactive hypoglycemia. The 3 commonest causes of reactive (or fed) hypoglycemia are alimentary, impaired glucose tolerance, and idiopathic. Salicylates, quinine, pentamidine, and disopyramide may lead to fasting hypoglycemia.

37) d.

Objective: review the factors that affect iron kinetics.

Serum ferritin is a measure of iron stores. It has very little day-to-day or diurnal variations. Its plasma levels are raised in acute phase responses and liver diseases. Transferrin saturation and serum iron are low during acute phase responses (similar to albumin; hence the designation negative acute phase reactants). Plasma iron is a measure of plasma iron availability. Transferrin levels are raised in pregnancy and with the use of oral contraceptive pills; low plasma levels are found in nephrotic syndrome, malnutrition, and liver diseases.

38) e.

Objective: review the risk assessment and severity of ulcerative colitis.

Assessing the severity of the disease will guide the management plan. In addition, it provides a prognostic outlook.

39) c.

Objective: review antigens/agents and their source(s) that are associated with hypersensitivity pneumonitis.

Cotton, flax, and hemp dust (textile industries) may result in byssinosis. Maple bark stripper's lung has been associated with exposure to Cryptostroma corticale (bark from stored maple). Aspergillus clavatus (moldy malting) is a cause of malt worker's lung, while exposure to avian proteins (excreta and feathers) may result in bird fancier's lung. Thermophilic actinomycetes exposure has been linked to the development of humidifier fever. Farmer's lung (the commonest one) has been linked to exposure to moldy hay and straw contaminated with Aspergillus fumigatus and micropolyspora faeni, while cheese worker's lung is associated with exposures to moldy cheese with Aspergillus clavatus and Penincillium casei.

40) d.

Objective: review the cardiovascular and hemodynamic effects (including effects on murmurs) of respiration and Valsalva's maneuver.

During inspiration, the JVP falls, the blood pressure falls, the heart rate accelerates, the right ventricular preload increases, and the second heart sound splits. The reverse occurs in expiration.

41) c.

Objective: review the role of random serum ACTH in differentiating various adrenal diseases.

Serum ACTH is elevated in primary adrenal failure (Addison's disease) because of loss of negative feedback inhibition on the pituitary, while it becomes low or undetectable in the secondary forms (e.g., hemochromatosis, brain trauma, brain surgery, pituitary irradiation, non-functioning pituitary tumors,…etc.). A random sample is also invaluable in evaluating the causes of Cushing's syndrome (ACTH-dependent versus ACTH-independent). The serum ACTH is not altered in polycystic ovarian syndrome.

42) d.

Objective: review the causes of aplastic anemia, and differentiate between aplastic anemia and pancytopenia.

Pancytopenia is the end result of many diseases acting on the bone marrow or peripherally. Megaloblastic anemia can result in pancytopenia, but the bone marrow is hypercellular. Aplastic anemia results in a hypo- or acellular marrow.

43) d.

Objective: review the types and associations of cryoglobulins and their clinical manifestations.

These antibodies, cryoglobulins, could be IgM or anti-IgG antibody complexes, which precipitate out in the cold (hence the initial designation as cryo-). Type I cryoglobulinemia encompasses monoclonal IgM antibodies which are usually associated with B-cell lymphoreticular malignancies (lymphoma, myeloma,…etc.). Type II disease (mixed or essential cryoglobulinemia) encompasses monoclonal IgM and polyclonal anti-IgG complexes. It has a rheumatoid factor activity; as a result, testing for rheumatoid factor may yield a very high titer.

It is mainly associated with hepatitis C infection (90% of cases) and may respond dramatically to interferon alpha and ribavirin (because the virus is complexed with these antibodies in the vasculitic lesions). Type II is also seen in B-cell malignancies and SLE. Type III disease refers to polyclonal IgM and anti-IgG antibodies, and is associated with autoimmune diseases (e.g., SLE) and chronic infections. By far the commonest type is type II; this produces purpuric rash over the lower limbs with peripheral neuropathy, arthralgia, and Reynaud's phenomenon.

44) e.

Objective: review Köebner's phenomenon and its significance.

This is the appearance of new lesions in areas subjected to skin trauma; this useful clue to the diagnosis in question.

45) e.

Objective: review organ specific and non-organ specific autoimmune diseases with respect to their associated HLA molecules.

HLA A3 has been linked to hereditary hemochromatosis.

46) b.

Objective: review the P-value

The p-value is the probability of obtaining a result equal to or more extreme than what was actually observed, assuming the null hypothesis. It measures the strength of evidence against the null hypothesis. One of the criticisms of the p-value is that it does not convey information in regard to the size of the observed effect. A small effect in a study with a large sample size can have the same p-value as a large effect in a small study. Another problem is that the more variables or endpoints in a study, the more likely one of them will come up statistically significant by chance alone, making the test's clinical significance circumspect. (Remember: a p-value of .05 means that, by chance alone, every 1 out of 20 times an event or finding will occur). Similarly, subgroup analysis, as well as repeated analysis of data during accrual with cessation of the experiment or trial when statistical significance is reached, is likely to lead to incorrect conclusions.

A type I error is committed by not accepting the null hypothesis when, in fact, it is true. Accepting the null hypothesis when, in fact, it is false is a type II error.

47) c.

Objective: review the medical treatment of hypertension during pregnancy.

Alpha methyldopa is the best medication studied during pregnancy; it acts as a central sympatholytic agent and has no effect on the uteroplacental circulation. Hydralazine is mainly used in the emergency management of pre-eclampsia or eclampsia. ACE inhibitors are contraindicated during pregnancy; they may result in abortion, fetal anomalies, neonatal renal failure, and fetal death. There is a debate and controversy about the use of diuretics during pregnancy because of the availability of safer and well-studied medications. If they were already prescribed "before" pregnancy, one can continue on them; however, don't initiate them during pregnancy. Nifedipine may also have a tocolytic effect; avoid its combination with magnesium sulfate because a dangerous hypotension may ensue. Of all beta blockers, labetalol is the safest one. It can be used in mild, moderate, and severe hypertension in pregnant women.

48) a.

Objective: differentiate between various desquamating skin disorders.

In Staphylococcal scalded skin syndrome (SSSS), there is an intra-epidermal line of cleavage but no inflammatory cell infiltration. In toxic epidermal necrolysis (TEN), the entire epidermis is involved and an inflammatory infiltrate is observed. In SSSS, there is involvement of the epidermis only and no mucosa affected while in TEN, there are involvement of mucous membranes and multiple organs as well. Nikolsky sign is positive in SSSS while it is negative in TEN. The skin lesions in SSSS heal with no scarring while the TEN lesions do heal with scar formation. The mortality rate is low in SSSS; a high mortality figure is seen with TEN, which is due to hypersensitivity reaction to drugs and medications.

49) e.

Objective: review the various types of amyloidosis, their causes, and their clinical manifestations.

Chronic tuberculosis, familial Mediterranean fever, juvenile rheumatoid arthritis, and Crohn's disease are causes of secondary "AA" amyloidosis. Primary "AL" amyloidosis is idiopathic (90% of cases) or is associated with plasma cell dyscrasia (10%).

50) d.

Objective: review the inclusion and exclusion criteria for the use of thrombolytic therapy in acute ischemic stroke.

The *inclusion* criteria for r-tPA infusion in ischemic stroke patients are: clinical diagnosis of ischemic stroke, onset of symptoms within three hours of the initiation of treatment (if the exact time of stroke onset is not known, it is defined as the last time the patient was known to be normal), and with a measurable neurologic deficit. The *exclusion* criteria are: stroke or head trauma within the prior 3 months, any prior history of intracranial hemorrhage, major surgery within 14 days, gastrointestinal or genitourinary bleeding within the previous 21 days, myocardial infarction in the prior 3 months, arterial puncture at a non-compressible site within 7 days, lumbar puncture within 7 days, rapidly improving stroke symptoms, minor and isolated neurologic signs, seizure at the onset of stroke with postictal residual neurologic impairments, symptoms suggestive of subarachnoid hemorrhage (even if the CT is normal), clinical presentation consistent with acute myocardial infarction (or post-myocardial infarction pericarditis), persistent systolic blood pressure >185 mmHg, (and/or diastolic blood pressure >110 mmHg, or requiring aggressive therapy to control this high blood pressure), pregnancy (or lactation), active bleeding, platelets count <100,000/mm3, serum glucose <50 mg/dL or >400 mg/dL, INR >1.7 (if on warfarin), prolonged aPTT (if on heparin), and abnormal brain scan (evidence of hemorrhage, evidence major early infarct signs, such as diffuse swelling of the affected hemisphere, parenchymal hypo-density, and/or effacement of >33% of the middle cerebral artery territory). Brain CT findings that preclude the use of rtPA are hemorrhage, large infarction, sulcal effacement, mass lesion, and prominent edema.

51) c.

Objective: review the paraneoplastic manifestations of various types of lung cancer and try to know their "clues".

The combination of this low blood urea level (syndrome of inappropriate secretion of ADH) and worsening cough in an ex-smoker man may well point out towards small-cell lung cancer. The latter can result in Lambert-Eaton myasthenic syndrome (fatigue and impotence). Squamous cell cancer produces hypercalcemia (PTH-related peptide secretion) and hypertrophic pulmonary osteoarthropathy (clubbing). Goodpasture syndrome may attack the lungs, kidneys, or both; however, it neither explains the impotence nor the bland urinary sediment. The normal blood counts would cast a doubt on polyarteritis nodosa (in which there is neutrophil leukocytosis).

52) e.

Objective: review high risk factors for sudden death in hypertrophic cardiomyopathy and how to prevent this event.

The normal blood pressure argues against this prominent degree of left ventricular hypertrophy (LVH). The presence of such potentially fatal cardiac dysrhythmia in the context of this prominent chamber enlargement, one should think of hypertrophic cardiomyopathy (HCM). This patient has 3 risk factors for sudden death, and implantation of cardioverter-defibrillator has been shown to improve survival. Amiodarone may lessen or even abort the dysrhythmia, but has no impact on survival. There is no evidence that she has an aberrant bundle to justify the use of radiofrequency ablation. "Wait and see" policy is totally inappropriate. She has no advanced heart block in order to place a permanent pacemaker. Risk factors for sudden death in HCM (the presence of at least 3 out of the following 6 parameters justifies the use of an implantable cardioverter-defibrillator):

1- Past history of failed cardiac death.
2- Recurrent syncope.
3- A strong family history of sudden death and/or an adverse genotype (e.g., troponin mutations).
4- Multiple attacks of non-sustained VT on 24-hour Holter monitoring.
5- Marked LVH.
6- Hypertensive response upon exercise.

53) b.

Objective: review the characteristic features of brain CT images in various intracranial hemorrhages.

The patient demonstrates the typical spider-leg sign of blood which has filled in the basal subarachnoid cisterns. This is acute spontaneous subarachnoid hemorrhage.

54) d.

Objective: review the causes and management of acute upper gastrointestinal bleeding.

Whatever the cause is, the first step should be the ABC (airway, breathing, and circulation). The patient should be hemodynamically stabilized before embarking on more specific therapies.

55) e.

Objective: review potentially reversible factors that affect renal function in uremic patients.

Many factors may result in sudden or rapid worsening of renal function in patients with well-stabilized uremia. These factors are hypertension or hypotension, urinary tract infection, urinary tract obstruction, nephrotoxic medications (commonly overlooked), and fever (systemic infections).

56) a.

Objective: review the protean features of acute HIV seroconversion illness and the methods to detect HIV infection during this time.

Oral ulcers occur in 20% of patients, while genital ulceration develops in 10% of cases. The syndrome mainly presents with fever, skin rash, fatigue, sore throat, and lymph node enlargement (mainly in the neck), 2 to 4 weeks after viral acquisition. Patients who develop severe or prolonged syndrome should be considered for receiving highly active anti-retroviral therapy (HAART).

57) d.

Objective: review antidotes in poisoning.

Hemodialysis should be strongly considered when the blood level of methanol exceeds 500 mg/dl.

The rationale of oral or intravenous ethanol therapy is to block the enzyme alcohol dehydrogenase, which metabolizes methanol to formic acid. The latter is responsible for this severe systemic acidosis and the high osmolal gap.

58) c.

Objective: review various cell types that are seen on peripheral blood smears and their clinical significance.

Bite and blister cells are seen in acute attacks of G6PD deficiency, and with the use of a supravital stain, Heinz bodies become obvious. Smudge cells are observed in CLL. Burr cells (echinocytes) are found in uremia, severe dehydration, burns, and old stored blood. Acanthocytes are seen alcoholic cirrhosis, pyruvate kinase deficiency, and abetalipoproteinemia.

59) b.

Objective: review diseases resulting from abnormalities in the complement system.

The late complement components (C5-C9) deficiency predisposes to recurrent Neisseria infections, while deficiency of the early complement system predisposes to autoimmune diseases (especially SLE). There is no C10 (a distraction).

60) d.

Objective: review drug-drug interactions in epileptic patients.

Apart from lamotrigine, the other options are enzyme-inducing agents and may produce contraceptive failure.

61) c.

Objective: differentiate between syncope and epilepsy.

The presence of repeated lapses of consciousness over this 2-year duration without injuries or further deterioration, one must think of non-epileptic attacks (psychogenic syncope). It is very unlikely that she a sinister cardiac cause!

62) d.

Objective: review the complications of long-standing rheumatoid arthritis.

Long-standing rheumatoid arthritis is a well-recognized cause of secondary AA amyloidosis; this usually targets the kidneys, ending up with nephrotic syndrome. Although subcutaneous fat aspiration and staining for Congo-red stain is reasonable to diagnose amyloidosis, but we should first confirm that nephrotic-range proteinuria is present at first.

63) e.

Objective: differentiate between various skin blistering disorders.

This is most likely pemphigus; the mouth is involved in 100% of cases. These oral lesions might be the presenting feature long before the skin lesions appear. The scalp and the natal cleft are one of the "hidden areas" of psoriasis.

64) d.

Objective: review drug-drug interactions as a cause of seizures in patients with chronic illnesses.

The metabolism of theophylline is already slow in elderly; the addition of an enzyme inhibitor (e.g., ciprofloxacin) would incite theophylline toxicity, which is the likely explanation for these seizures.

65) c.

Objective: review the causes of hyperprolactinemia and the levels of serum prolactin as a clue to the underlying cause.

Normally, serum prolactin does not exceed 500 mU/L. In non-pregnancy and non-lactation states, serum prolactin ranging from 500-1000 mU/L most likely results from medications or stress. Levels between 1000-5000 mU/L are usually due to medications, disconnection hyperprolactinemia, or microprolactinoma. Serum levels exceeding 5000 mU/L are indicative of macroprolactinoma. During pregnancy or lactation, serum prolactin may reach 20000 mU/L.

This patient's large pituitary tumor most likely represents a non-functioning macroadenoma, which has resulted in disconnection hyperprolactinemia. If the tumor was a macroprolactinoma, the serum prolactin must have reached a greater level.

66) c.

Objective: review the causes of renal artery stenosis and their consequences.

The presence of feeble pedal pulses in this elderly woman may well reflect a generalized process of atherosclerosis. The latter might quite well have involved the renal arteries. Fibromuscular dysplasia is seen the younger age group.

67) d.

Objective: review the causes of hyperuricemia and gout.

High-dose aspirin have a uricosuric effect, while a low-dose one results in uric acid retention and hyperuricemia. Pyrazinamide, thiazides, and alcohol impart a similar effect.

68) d.

Objective: review the various anatomical structures through a CSF needle passes.

When doing lumber puncture, your CSF needle will pass (in order) through the skin, superficial fascia, supraspinous and infraspinous ligaments, ligamentum flavum, areolar connective tissue (with the vertebral venous plexus), dura mater, and finally the subarachnoid matter. This distance ranges from about 2 cm in children to as much as 10 cm (or even more) in obese adults. The anterior longitudinal ligament lies more anteriorly.

69) b.

Objective: review types of thyroid cancer and the age group they target.

Papillary thyroid cancer is the commonest one (70% of all thyroid cancers) and is mainly seen in the young age group (less than 20 years of age). The follicular type usually develops in middle-aged people, while the anaplastic one occurs mainly in elderly.

70) b.

Objective: review the management of chronic fatigue syndrome and fibromyalgia.

The management of chronic fatigue syndrome does not differ a lot from that of fibromyalgia; graded exercises coupled with behavioral cognitive therapy and tricyclic antidepressants. Bed rest is not only without any benefit, but it is also counterproductive. The patient should be encouraged to undertake physical and mental exercises. Psychiatric counseling has been shown to produce a favorable response in 50% cases. Digging out for an organic cause is fruitless and will affect the patient's confidence; all patients should be assured that they have a disease that is neither fatal nor disfiguring. Although asking the patient to join a support group is a good advice, but stem "b" seems more reasonable at the time being.

71) e.

Review the medical management of atrial fibrillation.

Medical cardioversion can be successful in certain settings, and flecainide is a good choice. Flecainide is a class Ic anti-arrhythmic medication and is given orally as 100-200 mg, twice daily. It has a half-life of 20 hours and a hepatic route of elimination. Flecainide prolongs the PR, QRS, and QTc intervals. It is contraindicated in heart failure and can result in blurred vision, sinus dysfunction, and dizziness.

72) a.

Objective: review the lab findings in hemolytic uremic syndrome (HUS) and thrombotic thrombocytopenic purpura (TTP), and differentiate both of them from DIC.

A common mistake is to mix HUS (or TTP) with DIC. In HUS, the coagulation profile is normal apart from low platelets, i.e., normal PT, aPTT, TT, serum fibrinogen, and D-dimmer level. In DIC, these parameters are abnormal.

73) c.

Objective: review lithium toxicity.

Lithium is handled by the kidneys in a way that is similar to sodium. Any factor that interferes with kidney handling of this medication may result in its accumulation. Thiazides and ACE inhibitors are well-known culprits. Lithium has a narrow therapeutic margin; acute and chronic toxicities may ensue easily. Blood testing is diagnostic, and severe cases require hemodialysis. Fine tremor can be seen as a side-effect and does not reflect neurotoxicity; coarse tremor, seizures (may be multifocal), and coma are serious signs of toxicity.

74) b.

Objective: review the consequences of small bowel resection and their treatment.

Bile salt diarrhea results from the irritating effect of unabsorbed bile salts at the distal ileum, which act on the colonic mucosa; cholestyramine binds bile salts in the bowel lumen and improves diarrhea. Patient who are intolerant to cholestyramine may respond to aluminum hydroxide. The patient is at risk of vitamin B_{12} deficiency in the long-term; vitamin B_{12} injections are not needed for the time being.

75) c.

Objective: review auto-antibodies and their clinical correlations.

The constellation of CNS and skin involvement with vascular thrombosis at an unusual site (in this young woman) should always prompt you search for SLE or primary anti-phospholipid syndrome (APS). ANA is positive in 95% of SLE cases; this sensitivity is much lower in the primary APS.
Renal vein thrombosis and livedo reticularis favor APS over SLE (livedo reticularis occurs in 10% of SLE cases). Therefore, one should look for anti-phospholipid antibodies (or do lupus anticoagulant) to fulfill the diagnostic criteria for APS (i.e., one clinical feature + one lab feature).

76) d.

Objective: review the genetics of a1-antitrypsin deficiency.

Overt emphysema occurs only if the production of this anti-protease is <15% of the normal. There are many phenotypes, such as S, M, Z.

Only homozygous individuals for the Z allele (i.e., ZZ) develop the clinical consequences, because ZZ patients form less than 10-15% of this anti-protease. The disease has a "co-dominant" form of inheritance.

77) a.

Objective: review the causes of hyperparathyroidism.

The commonest cause of this raised serum calcium in outpatients is primary hyperparathyroidism, while the commonest cause of hypercalcemia in inpatients is malignancy (almost always from bone secondary tumors). Outpatients' hypercalcemia is usually mild and asymptomatic, and many cases do not need any intervention apart from regular follow-up.

78) b.

Objective: review the diagnosis and management of microalbuminuria in diabetes mellitus.

Microalbuminuria is a marker of incipient nephropathy in diabetics; ACE inhibitors have been shown to retard the progression, regardless of whether the blood pressure is elevated or not. Albuminuria is defined as:

1- Urinary albumin excretion of 30-300 mg/day (or 20-200 microgram/minute); this requires a timed urine collection, either 24 hours or overnight, *or*

2- Albumin:creatinine ratio (ACR) of >2.5 in males or >3.5 in females; normally it is 3-30 mg/mmol. This requires a random spot sample.

Either of the above findings should be reconfirmed at least twice over a 3-6 month period. For this patient, intensive blood glucose control is not required for the time being; achieving near euglycemia would suffice. However, if the patient is hypertensive, an aggressive lowering of his blood pressure is justified.

79) e.

Objective: review age-related ocular changes and differentiate them from pathological changes.

Down-beating nystagmus is never an expected age-related sign; it is always pathological and indicates a lesion at the level of the foramen magnum. Optokinetic nystagmus or few beats of nystagmus on an extreme sustained side-gaze are physiological. Age-related ptosis is the commonest cause of bilateral partial ptosis. The small pupils make fundoscopy difficult.

80) b.

Objective: review the diagnostic criteria and management of sick sinus syndrome (SSS).

In SSS, we may encounter a variety of dysrhythmia. The commonest ones are sinus bradycardia, SA arrest, supraventricular tachycardia, paroxysmal atrial fibrillation, and AV blocks. In symptomatic patients, the implantation of permanent pacemaker has been shown to improve bradycardia-associated symptoms (induced by SSS or related to drug therapy for tachyarrhythmia); however, this does not improve survival, and is not indicated in asymptomatic individuals.

81) c.
Objective: review various types of dermatitides.

This allergic contact dermatitis in this woman may well be due to paraphenylenediamine that is presents in hair dyes. The primary site of the rash is an important clue to the underlying contactant; for example, ear lobes in nickel allergy or the wrist in rubber glove allergy.

82) e.

Objective: differentiate between mania and schizophrenia, and bipolar disorder from unipolar major depression.

The boy does present with acute mania, but the overall clinical diagnosis is that of a bipolar disorder (note the past episodes of depression). Many patients are wrongly labeled as having schizophrenia; psychosis occurs in the depressive as well as the manic phase, and the delusions and hallucinations are usually consistent with the mood disturbance. All patients should be followed-up regularly, as stopping prophylactic therapy usually results in a relapse with a high risk of suicide.

83) c.

Objective: review the causes, mode of presentation, and management of acute COPD exacerbation.

Non-invasive ventilation for patients having a blood pH of 7.35-7.25 has been shown to reduce the need for endotracheal intubation, length of hospital stay, and inpatient mortality. When the patient deteriorates rapidly or the blood pH falls below 7.25 (i.e., profound systemic acidosis), invasive ventilation should be considered. However, when the outlook is poor because of co-morbidities (e.g., cancer, severe heart failure,…etc.), intravenous doxapram may be used.

84) b.

Objective: review the causes and clinical features of intracranial venous sinus thrombosis.

This female patient has a hypercoagulable state (post-partum) putting her at risk of thrombosis. In this case, the superior sagittal sinus was the target, which has resulted in headache, seizures, paraparesis, drowsiness, and papilledema. Brain CT scan would show parasagittal hemorrhagic infarctions. The intracranial pressure rises rapidly because of sudden impingement to the venous drainage. Subacute/chronic thromboses usually presents as pseudotumor cerebri. Watershed cerebral infarctions result from prolonged pan-cerebral ischemia, as in cardiac arrest or severe hypovolumic shock.

85) d.

Objective: review the clinical features and examination findings in complete heart block.

Complete heart block produces variable intensity of the 1st heart sound because of atrioventricular dissociation that results in variable filling of the ventricles.

86) a.

Objective: compare single blind, double blind, and triple blind studies.

Blinding occurs when the treatment assignment is hidden from the subject but not the observer.

Double-blinding means neither the observer nor the subject is aware of the treatment assignment. Triple-blinding means the observer, the subject, and the individual administering the treatment are not aware of the treatment assignments. The purpose of blinding is to minimize the possibility of observation bias in determining the outcome.

87) b.

Objective: review the management of acute pulmonary thromboembolism.

This post-partum woman has a high clinical probability for acute pulmonary thromboembolism. Those patients need acute anticoagulation while awaiting the results of documentary investigations. Subjecting the patient to series of lab and imaging testing in the hope to confirm or refute the diagnosis is unacceptable for the time being.

88) d.

Objective: review the side effects of bisphosphonates.

All bisphosphonates should be taken on an empty stomach with a full glass of water to enhance passage of the tablet and absorption. They are usually well-tolerated, but GIT upset may occur. Alendronate may cause esophagitis; therefore, it should be used with caution or to be avoided in patients with gastroesophageal reflux disease.

89) b.

Objective: review the stages of syphilis and their clinical manifestations and differential diagnoses.

The constellation of these features points out towards secondary syphilis; VDRL is used in the screening or to follow-up the treatment response. Hepatitis, whether clinical or subclinical, is a well-known complication of secondary syphilis.

90) d.

Review the management of recurrent renal calculi.

Thiazides reduce calcium excretion in urine; frusemide has the opposite effect. These renal calculi are not due to bone resorption process; therefore bone protecting agents are not indicated.

91) b.

Objective: review the organisms which cause endocarditis and their preferred target group.

This vagrant man might well be an intravenous addict; the clinical picture and lab findings are in favor of acute Staphylococcal endocarditis with possible lung showering. Streptococcus bovis endocarditis usually occurs in those with a GIT problem, usually colonic cancer in elderly people. "Early" prosthetic valve endocarditis is usually caused by Staphylococcus epidermidis.

92) c.

Objective: review the extra-glandular complications of primary Sjögren syndrome.

Distal renal tubular acidosis (RTA) is a well-characterized immune-mediated tubular complication of Sjögren syndrome. Hypercalciuria, nephrocalcinosis, and renal calculi all may occur. The distal RTA is usually "partial" without systemic acidosis. Sjögren syndrome has no specific treatment; only symptomatic measures.

93) b.

Objective: review the management of severe and life-threatening asthma.

The question did not mention any feature of life-threatening attack, such as coma, exhaustion, bradycardia, silent chest,...etc. Therefore, mechanical ventilation is not required. The patient is not responding to the conventional medical therapy; more frequent β_2 nebulization, adding nebulized ipratropium, or giving intravenous magnesium can be tried at this stage, provided that pneumothorax, bronchial plugging, or pulmonary collapse are not present. Prednisolone is given orally unless the patient has vomiting or inability to swallow; in the latter case, intravenous hydrocortisone is used.

94) a.

Objective: review the characteristic features of "idiopathic" trigeminal neuralgia.

Trigeminal neuralgia is not idiopathic if the patient is young, has bilateral or alternating attacks, or there are abnormal neurological signs (e.g., absent corneal reflex). In these cases, one should look for posterior fossa pathology, e.g., multiple sclerosis, pontine tumors,…etc.

95) b.

Objective: review the treatment of small-cell lung cancer versus non-small cancer.

Small-cell lung cancer could be limited (and treated by radiotherapy and chemotherapy) or extensive (treated by chemotherapy alone); there is no surgical option. Non-small lung cancer patients should undergo staging; the presence of T4, N3, or M1 would preclude surgery, as is the presence of FEV_1 <0.8 L and severe co-morbidities (e.g., severe heart failure, recent extensive myocardial infarction,…etc.). Age per se is not a contraindication.

96) d.

Objective: review Churg-Strauss vasculitis.

This woman demonstrates 4 out of the 6 diagnostic criteria for this type of vasculitis; asthma, paranasal sinus disease, mononeuritis multiplex, and eosinophilia. The remaining two criteria are fleeting chest opacities and eosinophilic vasculitis on sural nerve biopsy.

97) c.

Objective: review immunological tests for celiac disease.

Selective IgA deficiency co-exists in 2% of celiac patients; this renders serum testing for IgA anti-gliadin and anti-endomysial antibodies virtually useless. In addition, these tests are not quantitative; however, they are sensitive (85-95%) and highly specific (99%). In these cases testing for the IgG antibodies seems reasonable. Direct assays for tissue transglutaminase are even more accurate in selective IgA deficient patients. Note that IgA anti-gliadin and anti-endomysial antibodies may turn negative in successfully treated patients and they are usually negative in very young infants.

98) b.

Objective: review the side effect profile of antipsychotics.

Conventional antipsychotics can cause cholestatic jaundice; note the patient's itching. Although Wilson's disease targets individuals between the age of 5-55 years, it is unlikely in this patient; starting penicillamine without a clear-cut diagnosis is a wrong step. Drug-induced hemolysis is a possibility, but these are rare and are not necessarily Coomb's positive. "Wait and see" policy is not practical in a patient with unexplained jaundice. Serum antibodies against viral hepatitides may be negative early in the course, and repeating them at a later time seems reasonable but it is not the correct step for the time being. Stopping the culprit antipsychotic seems more acceptable.

99) c.

Objective: review factors that affect graft survival.

There are 4 types of graft reactions; hyperacute (within minutes to few hours and is caused by preformed serum antibodies), accelerated (within 2-5 days and is a cell-mediated one due to prior T cell sensitization to the donor antigens), acute (after 7-28 days which is principally cell-mediated but with some humeral component), and chronic (more than 3 months post-transplantation and has a variable combination of cell-mediated and humeral immunities). There is no delayed chronic phase.

100) b.

Objective: review causes of pes cavus and their clinical correlation.

Stork-like legs and pes cavus point out towards Charcot-Marie-Tooth disease. Some family members of this disease may show only pes cavus (forme fruste). Pes cavus may also observed in Friedreich's ataxia.

This page was intentionally left blank

Mock Paper Number Two

100 Best of Five Questions

This page was intentionally left blank

1) A 45-year-old woman presented with bilateral exophthalmos and fine postural tremor. You ran a battery of blood tests and finally diagnosed hyperthyroidism. All of the following are recognized features of hyperthyroidism, *except?*

   a. Resting tachycardia
   b. Apathy
   c. Anorexia
   d. Hirsutism
   e. Oligomenorrhea

2) A 24-year-old man visits the physician's office for a general check-up. You hear a cardiac murmur. The patient's 12-lead ECG and transthoracic echocardiography are unremarkable. Which one of the following is a characteristic feature of innocent cardiac flow murmurs?

   a. Harsh
   b. Usually mid-systolic
   c. Radiation to the lower neck
   d. Mainly heard at the cardiac apex
   e. Louder on squatting

3) A 39-year-old woman developed periodic cramping abdominal pains and diarrhea. She had asthma-like symptoms. After doing through investigations, the diagnosis turned out to be carcinoid syndrome. All of the following are skin manifestations of carcinoid syndrome, *except?*

   a. Hyperkeratosis
   b. Sclerodermatous changes
   c. Facial pallor
   d. Skin nodules
   e. Hyperpigmentation

4) A 29-year-old man was brought to the Emergency Room. His core body temperature is 40.3 °C. Which one of the following does *not* result in hyperthermia?

   a. Hypothyroidism
   b. Delirium tremens
   c. Aspirin poising

    d.   Amphetamine abuse
    e.   Generalized tetanus

5)   A 57-year-old man was diagnosed with polycythemia rubra vera. His ESR was 2 mm/hour. All of the following factors affect and subsequently lower the erythrocyte sedimentation rate (ERS), *except?*

    a.   Congestive heart failure
    b.   Chronic renal failure
    c.   Sickle-cell anemia
    d.   Hypofibrinoginemia
    e.   Cachexia

6)   A 45-year-old asks has surfaced the internet and read about metabolic syndrome and its clinical consequences. He asks if he has developed this syndrome. Which one of the following is a feature of metabolic syndrome?

    a.   Fasting blood glucose $\geq$110 mg/dl
    b.   Blood pressure $\geq$120/80 mmHg
    c.   Fasting serum triglyceride $\geq$ 250 mg/dl
    d.   Serum HDL-cholesterol < 30 mg/dl in women
    e.   Waist circumference >50 inches in men

7)   A young man was found to have hereditary angioedema because of deficiency in one of the complement system components. Which one of the following is not part of the classical complement pathway:

    a.   C1r
    b.   C1q
    c.   C2
    d.   Factor D
    e.   C4

8)   A 45-year-old man was brought to the Emergency Room because of drug-induced psychosis. Which one of the following substances and medications does not result in psychosis?

    a.   Alcohol
    b.   Isoniazid
    c.   Meperidine

d. Atenolol

e. Cocaine

9) A 51-year-old man complains of exertional shortness of breath. You intend to measure his diffusion capacity for carbon monoxide (DLCO). Which one of the following results in low DLCO with normal spirometry?

a. Polycythemia

b. Advanced emphysema

c. Right intra-cardiac shunting

d. During an asthmatic attack

e. Chronic pulmonary thromboembolism

10) A 54-year-old man had a blood pH reading of 7.5 because of a systemic illness. Causes of respiratory alkalosis:

a. Fever

b. Severe anemia

c. Rapid correction of metabolic acidosis

d. Myasthenia gravis crisis

e. Aspirin poisoning

11) A 32-year-old woman presented with diarrhea. You took a thorough history and did clinical examination. You think she has toxin-mediated diarrhea. Which one of the following organisms causes toxin-mediated diarrhea?

a. Campylobacter

b. Yersinia

c. Giardia

d. Bacillus cereus

e. Salmonella typhi

12) A young boy was diagnosed with Chédiak-Higashi syndrome. Which one is the correct statement about Chédiak-Higashi syndrome?

a. Can be autosomal recessive

b. There is diffuse skin hyperpigmentation

c. Defective eosinophil migration

d. Peripheral blood neutrophilia

e. May respond to vitamin A

13) A 68-year-old man presented with chronic constipation and bloody stools. You did colonoscopy and diagnosed colonic cancer. Which one of the following is not a poor prognostic factor in colorectal cancer?

a. Regional lymph node involvement
b. Bowel perforation
c. Absence of bowel obstruction
d. High pre-operative serum carcinoembryonic antigen level
e. Poorly-differentiated histology

14) An elderly male developed acute subdural hematoma after falling at home. Which one of the following is not a risk factor for falls in elderly people?

a. Previous history of falls
b. Wide corridor at home
c. Bilateral cataract
d. Cognitive impairment
e. Unsteady gait

15) A 56-year-old man was admitted to the Emergency Room because of focal neurological deficits. Your preliminary diagnosis is stroke. The list of the differential diagnosis of stroke includes all of the following, except?

a. Cervical radiculopathy
b. Seizures
c. Migraine
d. Conversion disorders
e. Hypoglycemia

16) A 65-year-old man has liver cirrhosis and cardiac dysrhythmia. You intend to prescribe a certain medications to control his heart rate. Which one of the following medications has hepatic route of elimination?

a. Sotalol
b. Amiodarone
c. Procainamide

    d.   Disopyramide
    e.   Dofetilide

17) A 70-year-old man presented with chronic constipation. You did barium enema and sigmoidoscopy. He had diverticular disease of the colon. Which one of the following is not a recognized complications of diverticular disease of the colon?

    a.   Hemorrhage
    b.   Fistula formation
    c.   Ileal cancer
    d.   Infection of the diverticula
    e.   Colonic stricture

18) A patient developed wide-spread purpura. His overall history and clinical work-up are suggestive of post-transfusion purpura. Which one of the following is the correct statement with respect to post-transfusion purpura?

    a.   Only occurs after whole blood transfusion
    b.   Mainly encountered in elderly males
    c.   Is diagnosed by demonstrating positive serum anti-PLA1 antibodies
    d.   Has poor response to intravenous immunoglobulin
    e.   Glucocorticoids are not effective in the treatment

19) A 43-year-old man has been diagnosed with AIDS. Today, he visited the physician's office because of hip pain. Boney abnormalities in HIV-infected individuals:

    a.   Avascular necrosis
    b.   Paget's disease of the bone
    c.   Osteosclerosis
    d.   Delayed metaphyseal closure
    e.   Increased risk of osteosarcoma

20) A young man developed delirium, motor neuropathy in both upper limbs, severe abdominal pain, and dark urine. Your preliminary diagnosis is acute intermittent porphyria. All of the following porphyrias match the corresponding enzyme deficiency, except?

a. Acute intermittent porphyria- aldehyde dehydrogenase
b. Porphyria cutanea tarda-uroporphyrinogen decarboxylase
c. Eryhropoietic porphyria-ferrochelatase
d. Hereditary coproporphyria-coproporphyrinogen oxidase
e. Variegate porphyria-protoporphyrinogen oxidase

21) A 72-year-old woman was diagnosed with giant cell arteritis. Which one of the following statements is the correct one with respect to giant cell (temporal) arteritis?

a. May co-exist with polymyalgia rheumatica
b. Headache is the least recognized symptom
c. Jaw claudication occurs in 90% of cases
d. The temporal artery is pulseless in a small number of patients
e. Responds favorably to low dose prednisolone

22) A 47-year-old vagrant man was brought to the Emergency Room. He was drowsy. You did some investigations and found that he had renal impairment, but you could not know whether it was acute or chronic renal impairment. All of the following features are in favor of chronic uremia rather than acute renal failure, except?

a. Osteitis fibrosa
b. Previous records of abnormal renal function
c. Small kidneys on ultrasound
d. Anemia
e. Presence of an AV shunt in the arm

23) A 54-year-old man had been experiencing progressive exertional breathlessness over the past 10 years. Eventually, he was found to have α1-antitrypsin deficiency. Which one of the following is the correct statement with respect to α1-antitrypsin deficiency?

a. Typically produces panacinar emphysema
b. Predilection for the apical lung areas
c. The resulting emphysema is symptomatic from the second decade
d. Chronic pancreatitis may occur
e. The deficient anti-protease is not commercially available

24) A 38-year-old female asks you about the pros and cons of oral contraceptive pills. Which one of the following is a recognized effect of long-term oral contraceptives administration?

   a. Reduction in serum triglyceride
   b. Improvement of depression
   c. Increased libido
   d. Breast tenderness
   e. Decreased appetite

25) A 54-year-old man has developed exertional retrosternal pain. He is hypertensive. You consider ischemic heart disease and intend to treadmill exercise test. Which one of the following is a contraindication to exercise ECG testing?

   a. Mild aortic stenosis
   b. Full thickness myocardial infarction within 4 weeks
   c. Chronic stable angina
   d. First degree heart block
   e. Acute aortic dissection

26) A 51-year-old man developed shortness of breath, persistent dry cough, and weight loss. His final diagnosis turned out to be lung cancer. He did not smoke cigarettes. Non-smokers who develop lung cancer might have exposed to all of the following, except?

   a. Passive cigarette smoke
   b. Chromium
   c. Lead
   d. Arsenic
   e. Cadmium

27) A 31-year-old man was recently diagnosed with seborrheic dermatitis because of scalp and forehead rash. Which one of the following is the correct statement with respect to seborrheic dermatitis?

   a. Is associated with absence of dandruff
   b. High skin sebum excretion is needed for the pathology
   c. Can cause otitis media
   d. Leads to scarring alopecia
   e. Might be confused with psoriasis

28) A 31-year-old man presented with hypertension, hypernatremia, and hypokalemia. His serum aldosterone is low and his plasma renin activity is also low. Which one of the following is a recognized cause of low serum aldosterone with low plasma renin activity?

    a.   Licorice abuse
    b.   Diuretic abuse
    c.   Renal artery stenosis
    d.   Conn's adenoma
    e.   Dexamethasone-responsive hyperaldosteronism

29) A man was diagnosed with a congenital long QT interval after developing recurrent attacks of palpitation. Which one of the following is the correct statement with respect to congenital long QT syndromes?

    a.   May be autosomal dominant
    b.   Usually become symptomatic in the 6th decade
    c.   The resulting palpitations are due to atrial tachyarrhythmia
    d.   Emotional stress lessens palpitation episodes
    e.   The ECG shows short PR interval

30) A young developed recurrent sino-pulmonary infections. You ordered a battery of investigations. Your final diagnosis is common variable immune deficiency syndrome. Which one of the following is the correct statement with respect to common variable immune deficiency syndrome?

    a.   Usually autosomal recessive
    b.   IgG is the only immunoglobulin that is deficient
    c.   The patient is protected against the development of autoimmune diseases
    d.   A differential diagnosis of cystic fibrosis
    e.   Most patients die during their second decade

31) A 47-year-old man has a long history of heartburn. You do esophagoscopy and find Barrett's esophagus. Which one of the following is the correct statement with respect to Barrett's esophagus?

    a.   A acute complication of gastroesophageal reflux disease
    b.   A risk factor for esophageal adenocarcinoma development
    c.   Should be treated by surgery whenever detected
    d.   Disappears on high dose omeprazole
    e.   Mainly targets the middle 1/3rd of the esophagus

32) A 21-year-old man underwent general check-up as part of his medical health insurance. His blood film reveals hypochromic anemia. Serum iron was lower than the normal reference range. Which one of the following conditions results in hypochromic anemia with low serum iron?

    a.   Sideroblastic anemia
    b.   Anemia of chronic diseases
    c.   Thalassemia trait
    d.   Vitamin B12 deficiency
    e.   Aplastic anemia

33) A 20-year-old man presented with myotonic features. You took a thorough history and did clinical examination. Your provisional diagnosis was myotonic dystrophy. Which one of the following is consistent with myotonic dystrophy?

    a.   Diplopia
    b.   Hypoglycemia
    c.   Bilateral cataract
    d.   Hirsutism
    e.   Low serum FSH

34) A woman was presented with dyspnea and cough. Her final diagnosis was lymphangioleiomyomatosis. Which one of the following statement is the correct one with respect to lymphangioleiomyomatosis?

    a.   Usually seen in postmenopausal women
    b.   Hemoptysis is against the diagnosis
    c.   Chylous pleural effusion occurs
    d.   Protective against the development of opneumothorax
    e.   Excellent response to high dose estrogen

35) A 17-year-old male presented with facial puffiness, abdominal distention, and pitting leg edema. Urine examination revealed proteinuria. Renal biopsy and histopathological examination showed minimal change glomerulopathy. Which one of the following is characteristically seen in minimal change glomerulopathy?

    a. Hypertension
    b. Active urinary sediment
    c. Non-selective proteinuria
    d. Poor response to glucocorticoids
    e. Normal renal function

36) A 54-year-old man was admitted because of acute atrial fibrillation. You intend to do give a medication in order to achieve a rhythm control. Which one of the following medications would you choose for pharmacologic cardioversion of atrial fibrillation?

    a. Carvedilol
    b. Digoxin
    c. Ibutilide
    d. Diltiazem
    e. Verapamil

37) A 54-year-old man was admitted to the Emergency Room because of pneumonic illness. His work-up has revealed Legionnaire's disease. Which one of the following statements is true about Legionnaire's disease?

    a. Caused by a Gram positive bacillus
    b. Abdominal pain and severe constipation are common
    c. There is cough with copious purulent sputum
    d. Results in raised serum transaminases levels
    e. Lowers serum creatinine kinase

38) A 22-year-old woman visits the Emergency Room having a panic attack. She has sweating, palpitation, and de-realization. Which one of the following is not a recognized feature of panic attacks?

    a. De-personalization
    b. Hypothermia
    c. Diffuse body trembling

d.   Chest pain
e.   Nausea

39) A 63-year-old woman visits the physician's office for a scheduled
check-up. She was diagnosed with rheumatoid arthritis few years ago.
You are looking for any extra-articular involvement. Which one of the
following is not a recognized extra-articular manifestation of
rheumatoid arthritis?

a.   Dry eyes
b.   Leg ulceration
c.   Spastic quadriparesis
d.   Aortic reflux
e.   Acute renal failure

40) A 61-year-old woman was diagnosed with type II diabetes 5 years ago.
She consults you today about her general condition. As part of her
medical examination, you do fundoscopy and find background diabetic
retinopathy. Which one of the following is a feature of background
diabetic retinopathy?

a.   Macroaneurysms
b.   Retinal detachment
c.   Retinal hemorrhage
d.   Vitreous hemorrhage
e.   Dot hemorrhages

41) A 20-year-old man was to found to have a hereditary
neurodegenerative disease after developing progressive instability of
stance and gait. His diagnosis was spinocerebellar ataxia type 1, due to
CAG trinucleotide repeat expansion. Which one of the following
disease is also associated with CAG trinucleotide repeats expansion?

a.   Friedreich's ataxia
b.   Myotonia dystrophica type I
c.   Huntington's disease
d.   Spinocerebellar ataxia type 8
e.   Fragile X chromosome syndrome

42) A 19-year-old man is brought to the Acute and Emergency department. He is a cocain addict. He demonstrate many features of cocaine intoxication. Which one of the following is a characteristic feature of cocaine intoxication?

    a. Intracerebral hemorrhage
    b. Calmness
    c. Hypotension
    d. Constricted pupils
    e. Hypothermia

43) A 32-year-old woman presented with recurrent upper abdominal colic. Ultrasound reveals gall stones. Laparoscopic cholecystectomy was done. The stones were of the pigment type. Which one of the following is recognized risk factors for the development of pigment gall stones?

    a. Female gender
    b. Ileal resection
    c. Obesity
    d. Clofibrate therapy
    e. Chronic hemolysis

44) A patient was found to have skin rashes on both dorsal aspects of the wrists. The appearance of the lesions was highly suggestive of lichen planus. Which one of the following is a feature of lichen planus?

    a. Usually occurs during adolescence
    b. Hepatitis C viral infection is an association
    c. The oral mucosa is protected from the disease
    d. Never involves the scalp
    e. The skin lesions are non-pruritic

45) A 67-year-old man was recently diagnosed with angina. He was hypertensive and diabetic, and smoked 2 packets of cigarettes per day. Features of typical chronic stable angina:

    a. Isolated shoulder pain is rare
    b. Usually the pain lasts less than 30 minutes
    c. The angina pain is stabbing in quality
    d. Relieved by prolonged rest

e.   Associated with hand tremor

46) A 45-year-old woman was admitted to the intensive care unit. She was diagnosed with glioblastoma multiforme 2 months ago. She demonstrated several signs and symptoms of raised intracranial pressure. Which one of the following measures can be used to control raised intracranial pressure?

a.   Controlled hypoventilation
b.   Oral mannitol
c.   Subarachnoid drainage
d.   Barbiturate coma
e.   Intravenous glucose water infusion

47) A 34-year-old man has been taking cyclosporine as mode of therapy for his severe and extensive plaque psoriasis. As part of his regular check-up, you find that serum magnesium is lower than the normal reference range. Which one of the following is not a recognized cause of hypomagnesemia?

a.   Cisplatin therapy
b.   Alcohol abuse
c.   Chronic diarrhea
d.   Prolonged hypoglycemia
e.   Thyrotoxicosis

48) Which one of the following factors is not a clue to a possible biologic attack in bioterrorism?

a.   The appearance of a large epidemic
b.   Appearance of a disease in an unusual area
c.   A rapid outbreak of a genetic disease in a community
d.   Appearance of a disease in the absence of its normal vector
e.   Concurrent epidemics of different kind of diseases

49) A 78-year-old man lives alone in his one-story house. His neighbor brings him today to the Emergency Room. The patient has confusion and dehydration. His blood sugar is more than 500 mg/dl. The social health worker visits his home and finds many sings of neglect. Which one of the following is not considered a sign of neglect of elderly?

a. Insufficient food supply
b. Unwashed dishes
c. Accumulated clothing
d. Old food in the frig
e. House in a rural area

50) A 70-year-old man is hospitalized for chest infection. Which one of the following factors favors delirium over dementia?

a. Slow onset of cognitive impairment
b. Prominent vegetative symptoms
c. Intact consciousness
d. Reversed sleep-wake cycle
e. Impaired memory

51) A 19-year-old athlete visits his doctor's office for a regular health check-up. He undergoes a battery of investigations. Which one of the following is not a normal finding in this man?

a. 4% eosinophils on peripheral blood counts
b. Few amorphous oxalate crystals on urine microscopy
c. FEV1/FVC ratio of 73%
d. 18 mm ventricular septal thickness on transthoracic echocardiography
e. Serum LDL cholesterol of 134 mg/dl

52) After a through skin examination of an HIV-positive patient, you detect multiple scattered bluish nodules. His recent CD4+ cells count was 130/ml3. What would you do next?

a. Examine the mouth
b. Start chemotherapy
c. Arrange for radiotherapy
d. Observe the lesions for spontaneous regression
e. Advise him to avoid contacting the skin of other people

53) A 45-year-old man has been recently diagnosed with chronic bronchitis. He tells you that he intends to quit cigarette smoking and asks for your help. All of the following are parts of smoke cessation programs, except?

a. Counseling
b. Nicotine skin patches
c. Engaging in a support group
d. Bupropion therapy
e. Nicotine inhalation

54) A 20-year-old female is doing her best to lose weight and to maintain it. She asks you if there is any medication that may help her lose more weight. Which one of the following does not result in weight loss?

a. Topiramate
b. Sibutramine
c. Phentermine
d. Orlistat
e. Sodium valproate

55) A 31-year-old man visits the Acute and Emergency Department saying that his right eye is red. You examine him and confirm this finding. All of the following are potential causes of red eye, except?

a. Viral conjunctivitis
b. Episcleritis
c. Chronic simple glaucoma
d. Anterior uveitis
e. Ocular trauma

56) A 65-year-old female presents with progressive abdominal swelling that turns out to be due to exudative ascites. Breast examination is unremarkable, as is its MRI. The fluid is sterile and contains many malignant-looking cells, but her abdominal CT scan fails to reveal any abnormality. Upper and lower GIT endoscopies were reported as being unremarkable. Her plain chest film does not look abnormal. Your final diagnosis is metastatic cancer with unknown primary. You plan to give chemotherapy for cancer of which one of the following organs?

a. Pancreas
b. Stomach
c. Ovary
d. Colon
e. Uterus

57) A 66-year-old man, who has long-standing poorly controlled hypertension, is brought to the Acute and Emergency Department with aphasia and right-sided weakness of 3 hours duration. The right planter reflex is extensor. Head CT scan reveals intracranial hemorrhage. Where is the site of the hemorrhage?

   a. Subarachnoid space
   b. Intracerebral
   c. Epidural
   d. Subdural
   e. Intraventricular

58) A 23-year-old female visits the physician's office because of malar skin rash. She reports arthralgia, pleuritic chest pain, and infrequent seizures. Her urinary sediment is active. What screening test would you choose for this woman?

   a. Anti-dsDNA antibodies
   b. Anti-histone antibodies
   c. Anti-nuclear antibodies
   d. Anti-Sm antibodies
   e. Rheumatoid factor

59) Because of prominent unintentional weight loss, a 17-year-old girl is brought by her mother to consult you. She has hypotension, cold extremities, lanugo hair over her back, and amenorrhea. She denies weight loss, and refuses to proceed with further testing. The most likely diagnosis is?

   a. An occult gastric cancer
   b. Anorexia nervosa
   c. Bulimia nervosa
   d. Physical/sexual abuse
   e. Agitated depression

60) A-54-year-old retired secretary has developed chronic myeloid leukemia and is due to receive imatinib mesylate as a mode of treatment. What is the mechanism of action of this medication?

a. Inhibiting the S-phase of mitosis
b. Inhibiting the M-phase of mitosis
c. Inhibiting tyrosine kinase
d. Inhibiting proto-oncogenes
e. Inhibiting normal apoptosis

61) A 12-year-old girl ingested 40 tablets of adult aspirin 5 hours ago. She has confusion and seizures. What is your best action for the time being?

a. Forced-alkaline diuresis
b. Gastric lavage
c. Hemodialysis
d. Activated charcoal
e. Give an antidote

62) After sustaining a femoral fracture from a road traffic accident, a young female is brought to Emergency Room with shock. Her blood urea and serum creatinine are rising. Her urinary sodium is expected to be?

a. More than 100 mEq/L
b. More than 20 mEq/L
c. Less than 10 mEq/L
d. More than 500 mEq/L
e. Less than 50 mEq/L

63) A 21-year-old male has frequent thumping heart beats which are a source of anxiety to him. Examination is unremarkable. Serum TSH is 3.1 mU/L. His ECG would probably show?
a. Multifocal ventricular ectopics
b. Atrial ectopics
c. Unifocal ventricular ectopics
d. Long QTc interval
e. Short PR interval

64) A 15-year-old school boy presents to the Emergency Room with mild lateral chest pain for 1 hour. His chest X-ray shows right-sided pneumothorax. After returning from the radiology department, he becomes dyspnic, tachypnic, and hypotensive, and finally collapses. What would you do next?

a. Another chest X-ray
b. Chest tube placement under water seal apparatus
c. Pleural aspiration using a wide-bore needle
d. Consult the thoracic surgery department
e. Give intravenous inotropic agents

65) A 69-year-old man is recovering from acute anterior wall myocardial infarction. You start secondary prophylaxis for this cardiac event. A statin is one of his medications. The goal of this statin is to achieve a serum LDL cholesterol of?

a. <200 mg/dl
b. <160 mg/dl
c. <130 mg/dl
d. <100 mg/dl
e. <50 mg/dl

66) A 17-year-old hemophiliac man from the Middle East was found to have elevated serum transaminases on routine blood testing. He denies jaundice, abdominal pain, or dark urine. You suspect chronic hepatitis C infection. The gold standard investigation in diagnosing this form of infection is?

a. Finding HCV RNA in the serum
b. Positive serum anti-HCV antibodies
c. Liver biopsy
d. Positive serum IgG anti-core antibodies
e. Elevated serum bilirubin

67) A 54-year-old female presents with diffuse bone pain and back pain 2 months after undergoing breast surgery for stage II invasive intra-ductal carcinoma of the right breast. Femoral X-ray film reveals an osteoblastic lesion. Biopsy from the lesion reveals estrogen receptor-negative but HER2-positive breast cancer metastases. In addition to combination chemotherapy, what would you add?

a. Anastrozole
b. Tamoxifen
c. Bilateral oöphorectomy
d. Leuprolide
e. Trastuzumab

68) A 7-year-old child presents with frequent short-lived staring and blinking which were noticed by his school teacher. His mother confirms the teacher's observation. The child denies such a thing. You examine the patient and find nothing remarkable. What does the boy have?

   a. Malingering
   b. Absence seizures
   c. Salaam attacks
   d. Pseudo-seizures
   e. Complex partial epilepsy

69) A 41-year-old woman with CREST syndrome has been complaining of exertional shortness of breath over the past 3 months. Chest examination is unremarkable and her chest X-ray reports normal findings. What complication you should look for in this woman?

   a. Recurrent pulmonary emboli
   b. Primary pulmonary hypertension
   c. Fibrosing alveolitis
   d. Recurrent aspiration
   e. Cardiac neurosis

70) A 60-year-old man presents with flaccid-roofed bullous lesions covering his body. His mouth has many red raw areas. A skin biopsy would most likely reveal?

   a. Intradermal line of cleavage
   b. Subdermal line of cleavage
   c. Hypodermal line of cleavage
   d. Deep fascial line of cleavage
   e. No line of cleavage

71) A 62-year-old female is referred to the surgical department after demonstrating a positive Pemberton's sign. This sign reflects the presence of thoracic outlet narrowing because of?

   a. Bilateral cervical ribs
   b. Multi-nodular goiter
   c. Thymic malignancy
   d. Large compressing aortic aneurysm

e. Chest Hodgkin's lymph nodes

72) A 12-year-old boy is referred to you having a mutation in chloride transporter in the ascending loop of Henle. He is likely to demonstrate all of the following, except?

a. Hypokalemia
b. Metabolic alkalosis
c. Hypercholesterolemia
d. Blood pressure of 90/50 mmHg
e. Increased urinary potassium loss

73) A 24-year-old man visits the physician's office because of recurrent facial and lip swelling. He states that sometimes he has respiratory distress and occasionally develops abdominal pain. He denies insect bites or urticarial lesions. What medication would you prescribe in the long-term?

a. Aspirin
b. Codeine
c. Morphine
d. Danazole
e. H1-blockers

74) A 34-year-old woman has severe fistulating Crohn's disease. She takes many daily medications. Which one of the following would probably be of benefit in treating her fistulae?

a. Mesalamine
b. Infliximab
c. Oral prednisolone
d. Metronidazole
e. Cyclosporine

75) A 32-year-old man is being admitted to the intensive care unit after developing overt respiratory failure. You consider endotracheal intubation with assisted ventilation. All of the following are invasive modes of assisted ventilation, except?

a. Pressure support ventilation
b. Controlled mandatory ventilation

    c.   Synchronized intermittent mandatory ventilation
    d.   Continuous positive airway pressure
    e.   Bilevel positive airway pressure

76) A 43-year-old homosexual HIV-positive man has been receiving highly active anti-retroviral therapy for 6 months. His has hyperlipidemia. Review of his pre-treatment serum lipid parameters shows borderline results. Which one of the following anti-retroviral agents is not responsible for this hyperlipidemia?

    a.   Zalcitabine
    b.   Indinavir
    c.   Nevirapine
    d.   Atazanavir
    e.   Lamivudine

77) A 5-year-old boy is found to have properdin deficiency after developing recurrent Neisseria meningitis. The mode of inheritance of this complement system regulatory protein is?

    a.   Autosomal co-dominant
    b.   Autosomal recessive
    c.   X-linked recessive
    d.   Sporadic
    e.   Autosomal dominant

78) A 50-year-old man was diagnosed with COPD last month. He has read a brochure published by a chest society about the management of COPD. The brochure mentions that certain vaccines are part of his illness management. You order influenza and the 23-polyvalent pneumococcal vaccines for him today. With respect to pneumococcal vaccine, when he should receive a booster dose?

    a.   After 6 months
    b.   After 1 year
    c.   After 3 years
    d.   After 5 years
    e.   After 10 years

79) A 34-year-old man was resuscitated in the Emergency Room after developing severe upper GIT hemorrhage, which was variceal in etiology. He has alcoholic cirrhosis. Which one of the following is used as part of a secondary prophylaxis against this form of bleeding?

   a.   Esophageal trans-section
   b.   Balloon tamponade
   c.   Octreotide
   d.   Propranolol
   e.   Isosorbide mononitrate

80) A 19-year-old man says he has felt a lump in the upper neck. Otherwise, the patient is healthy and denies any other compliant. His blood tests are within their normal reference range. You examine him and think this is a single lymph node. You order excisional biopsy. The histopathology reports toxoplasmosis. What would you do next?

   a.   Start pyrimethamine
   b.   Give spiramycin
   c.   Observe
   d.   Advise him to take sulfdiazine
   e.   Prescribe folinic acid

81) A 64-year-old man has been recently diagnosed with congestive heart failure. He visits you today to ask about his prognosis. Which one of the following factors portends a bad prognosis in this man?

   a.   Hypernatremia
   b.   Fourth heart sound
   c.   Low serum level of renin
   d.   Resting tachycardia
   e.   Normal blood pressure

82) A 40-year-old multi-parous woman presents with a sudden fall in hemoglobin and a tinge of jaundice, 1 week after receiving a pint of packed RBCs for severe menorrhagia. The serum indirect bilirubin is 3.0 mg/dl while the direct one is 0.4 mg/dl. There is raised serum LDH. Serum AST and ALT are normal. What has occurred in this woman?

a. Acute hemolytic transfusion reaction
b. Transfusion-associated CMV hepatitis
c. Delayed hemolytic transfusion reaction
d. Post-transfusion purpura
e. Reaction to sepsis

83) A 19-year-old female develops severe migraine attacks, mainly around menses. She is compliant with her amitriptyline tablets. These attacks are disabling and she is desperate for help. What would you prescribe next?

a. Low-dose estrogen during menstruation
b. Potassium tablets
c. Sumatriptan
d. High flow, high concentration nasal O2
e. Indomethacin

84) A 43-year-old female presents with central obesity, mooning of the face, skin striae, and hypertension. The commonest cause of Cushing's syndrome is?

a. Pituitary adenoma
b. Exogenous glucocorticoids administration
c. Benign adrenal tumors
d. Small-cell lung cancer
e. Adrenal carcinoma

85) Many members of a family were found to have hyperglycemia and their final diagnosis is maturity onset diabetes of the young type 3. This subtype is due to mutations in?

a. Hepatic nuclear factor 1-α
b. Hepatic nuclear factor 4- α
c. Glucokinase gene
d. Hepatic nuclear factor-γ
e. Insulin receptor

86) A 32-year-old woman checks her blood pressure regularly. Her values range between 120 to129 mm Hg systolic and 80 to 89 mmHg diastolic. According to blood pressure categorization, what category this woman's hypertension belongs to?

a.   Stage II hypertension
b.   Normal blood pressure
c.   High normal blood pressure
d.   Stage I hypertension
e.   Stage III hypertension

87) A 30-year-old man has developed end-stage renal disease because of cystinuria. What is the best treatment modality?

a.   Alkalization of urine
b.   Penicillamine
c.   Renal transplantation
d.   Captopril
e.   Increasing fluid intake

88) A 65-year-old man presents with epistaxis, shortness of breath, cough, and rapidly progressive glomerulonephritis. His serum cANCA is positive. This antibody attacks which one of the following?

a.   Neutrophil serine protease-3
b.   Neutrophil myeloperoxidase
c.   Nuclear histone
d.   Mitochondria
e.   Lysosomes

89) Because of fever, weight loss, and cough, a 49-year-old man visits your office. He also reports dyspnea and excessive sweating. He has been experiencing these symptoms for 3 weeks. He has failed to respond to oral clarithromycin. Chest X-ray reports a photographic negative pattern of pulmonary edema. What investigation you plan to do?

a.   Complete blood counts
b.   Chest CT scan
c.   Stool for Ascaris
d.   Bronchoscopy
e.   Sputum cytology

90) A 54-year-old high school teacher has been experiencing abdominal pain, steatorrhea, and weight loss over the past several months. The plain abdominal film shows flecks of pancreatic calcification. What is the commonest cause of chronic pancreatitis in the Western World?

a. Crytpogenic
b. Gall bladder disease
c. Alcohol
d. Pancreatic tumors
e. Autoimmune

91) A 55-year-old man presents with resistant hypertension and hypercalciuria. His serum calcium is 12.0 mg/dl. Your preliminary diagnosis is primary hyperparathyroidism. How would you treat?

a. Parathyroidectomy
b. Pamidronate
c. Calcitonin
d. Mithramycin
e. Fluid resuscitation

92) A 64-year-old man has asymmetric resting tremor, bradykinesia, and festinant gait. He demonstrates a favorable response to L-dopa therapy. With regard to this man's disease, choose the correct statement?

a. The diagnosis is mainly clinical
b. Brain MRI should be done in all cases
c. Brain CT with contrast aids the diagnosis
d. Toxicology screen is synergistic
e. SPECT and PET scans are very helpful in the diagnosis and follow-up

93) A 33-year-old female develops left-sided posterior tibial artery thrombosis, one week after starting anticoagulation therapy for right-sided deep venous thrombosis of the leg. The aPTT is 2.5 times the control and the INR is 2. Which test you intend to order to confirm your clinical diagnosis?

a. Repeat the INR
b. Re-order aPTT
c. Complete blood counts
d. Factor V Leiden testing
e. ANA

94) A 25-year-old woman visits the physician's office because she has surfed the internet and found that women with BRCA1 genetic mutations are at risk of developing malignancies apart from breast cancer. Her older sister died of BRCA1-associated breast cancer at the age of 30 years. The woman is positive for this mutation. In addition to breast cancer, she is at risk of developing which one of the following?

  a. Small-cell lung cancer
  b. Skin basal cell cancer
  c. Ovarian epithelial cancer
  d. Cervical squamous-cell cancer
  e. Acute myeloblastic leukemia

95) A 69-year-old man comes for an annual check-up visit. You hear a grade II/VI ejection systolic murmur at the right upper sternum. Otherwise, he is healthy and enjoys an independent life. He does the daily shopping and denies any chest symptoms. Echocardiography shows mild aortic stenosis. What would you do for him?

  a. Observe
  b. Arrange for coronary catheterization
  c. Repeat the echo study after 3 months
  d. Give aspirin
  e. Advise for aortic valve replacement

96) A 34-year-old Swedish man is on a vacation trip in Malaysia. He has developed around 6 watery bowel motions per day. There are no blood or pus in stool. Which one of the following is the commonest cause of travelers' diarrhea?

  a. Enterohemorrhagic E. coli
  b. Enterotoxigenic E. coli
  c. Enteroinvasive E. coli
  d. Non-typhoidal Salmonella
  e. Campylobacter jejuni

97) A 28-year-old pregnant woman, in her mid-trimester, is referred to you from the obstetric clinic because of an abnormal lab test. Which one of the following results would be abnormal in this woman?

a. One centimeter increase in kidney length
b. Serum bicarbonate of 23 mEq/L
c. Serum creatinine of 1.5 mg/dl
d. PaCO2 of 30 mmHg
e. 200 mg/day urinary protein

98) A 75-year-old female presents with morning stiffness, skin nodules, and feet numbness. Examination reveals ulnar deviation and peripheral neuropathy. You diagnose rheumatoid arthritis. Part of your management plan is to protect this patient against NSAID-induced gastric ulceration. Which one of the following is not used in this prophylactic program?

a. Lansoprazol
b. Ranitidine
c. Misoprostol
d. Sodium aliginate
e. Use of COX-2 inhibitor instead of NSAIDs

99) You read an original article that was published in a medical journal about a new screening test for SLE called lupin-1 antibodies, and the results were shown as follows:

	SLE patients	Individuals with no SLE
Positive result	96	16
Negative result	4	84

What is the positive predictive value of this novel testing?

a. 84%
b. 99%
c. 95%
d. 80%
e. 90%

100)   A 54-year-old male visits the Emergency Room having had fever, dyspnea, and increased cough and sputum purulence over the past 2 days. He has been diagnosed with COPD two years ago. He is in respiratory distress and you consider starting CPAP. Which one of the following is a contraindication to the use of CPAP?

    a.   The patient is cooperative
    b.   Synchronous breathing is present
    c.   The patient can protect airways
    d.   Glasgow coma scale of less than 10
    e.   The patient has slight airway secretions

Mock Paper Number Two

Answers

This page was intentionally left blank

1) d.

Objective: review the atypical features of hypothyroidism and hyperthyroidism.

Heart failure, leg edema, exacerbation of asthma all may be caused by hyperthyroidism. The apathetic form of hyperthyroidism is mainly seen in elderly. The incidence of anorexia increases with the age of the patient. Unexplained weight loss may dominate the picture in elderly people, wrongly suggesting an occult malignancy. Diffuse scalp hair thinning occurs; hypertrichosis is seen in hypothyroidism. Menorrhagia is mainly observed in middle-aged hypothyroid women; oligomenorrhea or amenorrhea occurs in hyperthyroidism.

2) b.

Objective: review the characteristics of cardiac murmurs.

Flow cardiac murmurs tend to be soft, sometimes with an ejection quality. They are systolic, never diastolic in timing. There is no or variable radiation and they are mainly heard at the cardiac base. Maneuvers, such as standing or hand grip, have no effect on these murmurs.

3) c.

Objective: review the skin manifestations of systemic diseases.

Facial flushing with cramping abdominal pain, diarrhea, and wheezes occur in the acute episodes of carcinoid syndrome. Chronic cases of carcinoid syndromes may show skin changes indistinguishable from those of scleroderma while others may show hyperpigmentation and hyperkeratosis skin changes resembling pellagra. The presence of skin or subcutaneous nodules in carcinoid syndrome patient most likely represents metastasis, although this is rare in clinical practice.

4) a.

Objective: review the causes of hyperthermia and differentiate between hyperthermia and fever.

Hyperthermia may result from hyperthyroid storm, pheochromocytoma, malignant hyperthermia, neuroleptic malignant syndrome, hypothalamic tumors and infiltration, heat stroke, encephalitis, cocaine abuse, anti-cholinergic poisoning, occlusive dressings, status epilepticus, and dysautonomia.

5) b.

Objective: review factors that influence the ESR.

The ESR becomes low in sickle cell anemia, polycythemia, extreme leukocytosis, altered red cell shape (e.g. acanthosis), hypofibrinoginemia, and congestive heart failure. The ESR rises normally with age and during pregnancy. Chronic renal failure, hypercholesterolemia, and plasma cell disorders elevate the ESR.

6) a.

Objective: review the criteria of metabolic syndrome.

The presence of at least 3 out of the following 5 variables defines metabolic syndrome:

1- Blood pressure ≥130/85 mm Hg.
2- Fasting blood glucose≥110 mg/dl.
3- Abdominal obesity: waist circumference >35 inches in women and >40 inches in men.
4- Serum HDL cholesterol < 50 mg/dl in women and <40 mg/dl in men.
5- Serum triglyceride≥150 mg/dl.

7) d.

Objective: review the complement pathways.

C1q, C2, and C4 are parts of the classical complement system. C5 is belongs to the membrane attack complex (terminal pathway). C1q is a fraction of C1 complex and binds Fc portion of IgG and IgM; C1s and C1r are parts of the same complex. C2 is a protease which cleaves C3 and C5. Factor D is a regulatory protease of the alternative pathway which cleaves factor B.

8) d.

Objective: review drug-induced psychosis.

Prominent psychosis may appear as part of alcoholic hallucinosis and during withdrawal states. The anti-TB drugs, isoniazid and cycloserine, may result in psychosis. The toxic metabolites of meperidine may incite a psychotic reaction. Amiodarone and digoxin uncommonly produce psychosis. Cocaine, especially the crack one, is another culprit.

9) e.

Objective: review factors that affect DLCO and KCO.

Anemia, early cases of emphysema (before any clinically significant airflow obstruction appears), primary pulmonary hypertension, and interstitial lung fibrosis (before any reduction in lung volumes) may decrease DLCO but the spirometry is normal. During asthma attacks (obstructive patter on spirometry), the DLCO is usually normal; rarely, it may increase by an unknown mechanism. Factors that increase the DLCO are polycythemia, right-sided intracardiac shunts, alveolar hemorrhage, and increased pulmonary blood flow during exercise.

10) d.

Objective: review acid/base disturbances.

Rapid breathing causes a washout of CO_2 with resultant respiratory alkalosis. This may be produced by fever (direct stimulation of the medullary respiratory center), severe anemia (due to hypoxemia) and rapid correction of metabolic acidosis. Myasthenic crisis and post-polio syndromes result in hypoventilation and respiratory acidosis with CO_2 retention.

11) d.

Objective: review diarrhea mechanisms.

Campylobacter is an invasive pathogen that results in mucosal inflammation with watery and/or bloody diarrhea. Yersinia is another invasive pathogen which may cause acute diarrhea or Crohn's disease-like chronic ileo-colitis. Giardia directly contacts the enteric mucosa and this can result in watery diarrhea, malabsorption, and dyspepsia. Bacillus cereus secretes an exotoxin which causes gastroenteritis. Clostridium difficile results in the release of cytopathic toxin which may end up with toxic megacolon. Salmonella typhi is an invasive pathogen.

12) a.

Objective: review the inherited white blood cell disorders.

Chédiak-Higashi syndrome is an autosomal recessive disease of defective neutrophil migration and bacterial killing; sometimes it is sporadic. Recurrent infections are very common and patients have partial albinism. Neutropenia and thrombocytopenia develop and some patients may display a partial response (and improvement) to vitamin C administration.

13) c.

Objective: review the poor prognostic factors in common malignancies.

A gloomy outlook is portended if patients with colorectal cancer have developed regional lymph node involvement (especially if more than 4 nodes were involved), bowel perforation (and tumor adherence to adjacent organs), bowel obstruction, allelic loss of chromosome 18q, pre-operative CEA serum level of more than 5 ng/ml, and a poorly differentiated histology.

14) b.

Objective: review risk factors for recurrent falls and their consequences.

A previous history of fall creates a sense of unsteadiness and fear of future falls. Females, especially those who have low body mass index, unsteady gait, and general frailty are at risk of falls. Home factors, such as narrow corridor, slippery floor, things scattered on floor,…etc., should not be overlooked.

15) a.

Objective: review the stroke-like presentation of neurological and non-neurological diseases.

Brain tumors, whether primary or secondary (especially hemorrhagic tumors) may present acutely. Post-ictal Todd's paresis may convey a wrong impression of stroke. Complicated migraine (with focal or lateralizing signs) may be have a stroke-like presentation; it is a diagnosis of exclusion. Conversion disorders and malingering should always be considered when the neurological signs are inconsistent and loose. Hyperglycemia and hypoglycemia may result in focal signs, seizures, confusion,…etc.; the signs may fluctuate and/or alternate between the 2 sides.

16) b.

Objective: review the metabolism of anti-arrhythmic medications and their precautions in organ failure.

Generally, 4 medications have a renal route of excretion, and these are sotalol, dofetilide, procainamide, and disopyramide. Amiodarone has a hepatic route of elimination.

17) c.

Objective: review the complications of diverticular disease of the colon and their management.

Diverticular hemorrhage is usually sudden and painless and stops spontaneously in 70% of cases; overall, it is uncommon. Fistula formation may occur, usually colo-vesical; cutaneous fistulae are rare and usually develop after surgery. Note that the commonest causes of colo-vesical fistulae (in order of frequency) are diverticular disease of the colon, colonic cancer, and Crohn's disease. Ileal cancer is not an association with diverticular disease of the colon; however, it may increase the risk of left-sided colonic cancer. Colonic stricture may be acute because of colonic wall edema or secondary ileus, or less commonly, chronic which may resemble malignancy. Note that both conditions may co-exist in the same patient.

18) c.

Objective: review post-transfusion purpura and its treatment.

Post-transfusion purpura is usually seen 2-10 days after receiving whole blood, packed RCBs, or any blood product that may contain platelets. The usual victims are multi-parous women. The diagnosis is secured by demonstrating positive serum anti-PLA1 antibodies; the patient is usually homozygous for the PLA2 genotype. Favorable response to intravenous immunoglobulin is observed in 80% of cases; plasma exchange has the same efficiency, but is usually inconvenient for most patients. Steroids may produce improvement, but high doses are usually required (with many side effects); generally, they are less effective than intravenous immunoglobulin therapy.

19) a.

Objective: review the bone abnormalities in HIV infection.

HIV avascular necrosis usually targets the knees and hips and is frequently bilateral and multifocal. Osteopenia and osteoprosis are being recognized increasingly with no single explanation behind them. HIV patients are not at increased risk of developing osteosarcomas.

20) a.

Objective: review the metabolic pathway of hemoglobin synthesis and porphyria defects.

You should remember at least the mutated enzymes of acute intermittent porphyria (prophobilinogen deaminase) and porphyria cutanea tarda (uroporphyrinogen decarboxylase) because both are common examination themes.

21) a.

Objective: review temporal arteritis and compare it with polymyalgia rheumatica.

About half of temporal arteritis patients have polymyalgia rheumatica, and 15% of polymyalgia rheumatica patients have temporal arteritis. Headache is the commonest symptom, usually throbbing and severe. Jaw claudication is usually intermittent and seen in one third of cases. Temporal artery pulselessness and/or tenderness are observed in 50% of cases only. Steroids should be used in high doses, usually 60-80 mg of prednisolone per day. The response is dramatic within 24 hours in the majority.

22) d.

Objective: differentiate between acute and chronic renal failure.

Bone disease, as evidenced by abnormal X-rays, is a strong evidence against acute renal failure. Boney changes needs long time to establish themselves. Previous documentation of abnormal renal function suggests that the renal failure is already established and not an acute process. Normal or increased kidney size can be encountered in both, acute and chronic renal impairment; however, small contracted kidneys favor uremia over acute renal failure. Anemia is common in renal impairment, whether acute or chronic. Any evidence of "chronic" renal replacement therapy (cubital fossa AV shunt, peritoneal Tenkoff's catheter, scar of renal transplant) points out towards an underlying uremia.

23) a.

Objective: review the genetics and features of a1-antitrypsin deficiency.

Typically patients develop panacinar emphysema, which is more pronounced at the lower lung zones (because of increased neutrophil load in the lower zones). However, smokers may have chronic bronchitic changes as well. Non-smokers usually become symptomatic with dyspnea around the age of 40-50 years; smokers present earlier by an average of 10 years. Liver cirrhosis is seen in 10% of adults; children and infants have a higher incidence of liver dysfunction (around 20%). Purified α1-antitrypsin can be given as weekly infusions. The patient should be advised to quit cigarette smoking and to avoid passive smoking situations. Single lung transplantation improves lung function and survival.

24) d.

Objective: review the side effect profile of long-term oral contraceptives (OCPs) administration.

OCPs increase serum triglyceride and HDL cholesterol and impart an anti-insulin action. They may result in exacerbation of depression and induction of fatigue in some patients. Decreased libido is noted by few females. OCPs are a risk factor for gall stones formation as well as impairment of blood pressure and diabetes control. Breast tenderness is common, and reduction in vaginal secretions may occur. The appetite increases and weight gain and bloating are common.

25) e.

Objective: review the contraindications and limitations of exercise ECG.

The contraindications for doing exercise ECG testing are severe symptomatic aortic stenosis (or any severe left ventricular out flow obstruction), recent myocardial infarction within 48 hours, unstable angina (that has not been stabilized), decompensated heart failure, uncontrolled hemodynamically significant arrhythmias, and pulmonary embolism/infarction. Stenosis of the left coronary artery main stem is a *relative* contraindication, as are moderately stenotic valvular lesions, severe uncontrolled hypertension, hypertrophic cardiomyopathy, and high-degree AV block.

26) c.

Objective: review risk factors for lung cancer.

The other risk factors are radiation, asbestos, and chromoethyl ether. Lead is not a risk factor for lung cancer.

27) e.

Objective: review the general features of seborrheic dermatitis.

Seborrheic dermatitis is associated with excessive dandruff formation. Sebum excretion may be low, normal, or high; therefore, seborrhea is not essential for the pathogenesis. Seborrheic dermatitis is the commonest cause of otitis externa; it may also result in blepharitis and conjunctivitis. It does not produce hair loss, but severe intervening infections might end up with scarring alopecia. Seborrheic dermatitis is a differential diagnosis of psoriasis and other chronic papulosquamous eczematous skin disorders.

28) a.

Objective: differentiate between primary aldosteronism, secondary aldosteronism, and apparent aldosteronism.

Patients with low serum aldosterone and low plasma renin activity have "apparent aldosteronism or pseudo-hyperaldosteronism"; they demonstrate hypertension and hypokalemia. The usual causes are licorice abuse (usually in the form of chewing tobacco), Liddle's syndrome, 11β-hydroxysteroid dehydrogenase deficiency, 11-deoxycotricosteron secreting tumors, and 17-hydroxylase deficiency. Diuretic abuse and renal artery stenosis result in secondary aldosteronism with high serum aldosterone and plasma renin activity. Conn's adenoma and dexamethasone-responsive hyperaldosteronism are the causes of primary aldosteronism with high serum aldosterone and low plasma renin activity.

29) a.

Objective: review congenital long QT interval syndromes and their management.

Inherited syndromes of long QT syndrome are Jervell-Lange-Nielsen syndrome (which is autosomal recessive and is associated with sensori-neural deafness; its course is highly malignant) and Romano-Ward syndrome (which is autosomal dominant and is a pure cardiac phenotype; its course is generally benign when compared with Jervell-Lange-Nielsen). Most cases result from mutations in various components of the cardiac *potassium* channels causing delayed phase 3 of cardiac action potential; only LQT3 subtype results from *sodium* channel mutation causing prolongation of the phase 2 of cardiac action potential. The symptomatology usually starts in adolescence; sudden death may be the presentation in some. Ventricular, not atrial, tachyarrhythmia occurs. Emotional stress as well as undue exertion may trigger attacks of ventricular tachycardia. The only consistent ECG feature is long QT interval.

30) d.

Objective: review immunoglobulin deficiency states.

Common variable immune deficiency syndrome is a sporadic disease and all immunoglobulins are affected variably. It is associated with several autoimmune diseases; pernicious anemia ranks first on this list. Young patients with recurrent sino-pulmonary disease and bronchiectasis may be misdiagnosed as cystic fibrosis. The prognosis is excellent if patients receive optimal management (vigorous use of antibiotics for infections and regular intravenous immunoglobulin administration). The life span is usually normal.

31) b.

Objective: review Barrett's esophagus and its treatment options.

Barrett's esophagus is a chronic complication of long-term exposure to gastric acid; it is asymptomatic but confers and increased risk for future development of esophageal adenocarcinoma. Although prolonged courses of high doses of proton pump inhibitors may result in partial response in term of disappearance of the metaplastic tissue; the latter does not completely disappear and the risk of malignancy stays. The lower 1/3rd of the esophagus is involved for a variable distance; extension for more than 3 cm implies a long segment involvement and a severe reflux disease.

32) b.

Objective: review the differential diagnosis of hypochromic anemia.

In sideroblastic anemia, the serum iron and transferrin saturation are both raised. The anemia of chronic diseases may be hypochromic in 25% of cases; both, serum iron and total iron binding capacity are low, implying a defect in iron kinetics. In thalassemia minor, the serum iron is normal and there are normal iron binding capacity and transferrin saturation; however, microcytosis is usually prominent and is disproportionate to the mild hypochromia. Iron deficiency anemia is the prototype of hypochromic anemia; serum iron is low with a low transferrin saturation but increased total iron binding capacity. The anemia of aplastic anemia is not hypochromic; it is normochromic normocytic and at times displays an increased MCV. Vitamin B_{12} deficiency results in megaloblastosis.

33) d.

Objective: review dystrophic myotonia.

Myotonia dystrophica attacks the eyelids producing bilateral ptosis without frontalis over-action; the extra-ocular muscles are preserved and no double vision occurs. Overt hyperglycemia or impaired glucose tolerance test occur in 10-20% of patients. One of the earliest features of the disease is the development of bilateral stellate cataract. Dysphagia and dysarthria reflect pharyngeal and laryngeal myotonia. Hypergonadotrophic hypogonadism (raised serum LH and FSH) is a secondary feature of testicular atrophy. Men usually develop early frontal baldness.

34) c.

Objective: review interstitial lung diseases and their characteristics.

Lymphangioleiomyomatosis characteristically attacks women during their child-bearing age. Therefore, estrogen is thought to play a role in the pathogenesis; however, estrogen ablation therapy is questionable. Hemoptysis, dyspnea, and cough may occur. Chylous pleural effusions occur with thoracic duct dilatation. Recurrent pneumothorax is a risk and results from rupture of the underlying lung cysts.

35) e.

Objective: review the characteristics of glomerulopathies.

In minimal change glomerulopathy, the blood pressure is normal but it may rise after starting high dose prednisolone. Apart from containing excess protein and fat globules, urine examination is unremarkable with a bland sediment; an active sediment with red cell casts indicates a nephritic process. The proteinuria is highly *selective*; non-selective proteinuria occurs in other glomerulopathies and implies a poor response to immune suppressants. Most patients show a favorable response to high dose prednisolone. Progression towards chronic renal failure does not occur.

36) c.

Objective: review the management of atrial fibrillation.

Propranolol (and other beta blockers), verapamil, digoxin, and diltiazem are useful for rate control (i.e., slowing the ventricular response). Flecainide, ibutilide, dofetilide, quinidine, and propafenone are used in pharmacologic cardioversion and maintenance of sinus rhythm in atrial fibrillation.

37) d.

Objective: review Legionnaire's disease.

Legionnaire's disease is caused by an aerobic Gram-negative bacillus, L. pneumophila. Extra-thoracic manifestations (diarrhea, confusion,…etc.) may dominate the clinical picture in some patients. The resulting cough is non-productive at the beginning, and then it becomes productive of scanty sputum (which is usually non-purulent). Hepatic involvement with raised serum AST and ALT are common. Proteinuria and microscopic hematuria may be seen; acute renal failure may develop. The serum CPK is raised; rhabdomyolysis may ensue.

38) b.

Objective: review generalized anxiety disorders.

During a panic attack, fear of dying and fear of going crazy are prominent. There may be de-realization, de-personalization, and fainting. The whole body may tremble with palpitation, shortness of breath, and chocking sensation. Nausea and abdominal discomfort are well-recognized features.

39) e.

Extra-articular manifestations of rheumatoid arthritis may include dry eyes (as part of secondary Sjögren's syndrome), leg ulceration (or frank gangrene, from vasculitis), spastic quadriparesis (atlanto-axial subluxation or subluxations at a subaxial level with secondary spinal cord compression), aortic regurgitation, and secondary "AA" amyloidosis (with nephrotic syndrome). Acute renal failure is not a sequela.

40) e.

Objective: review diabetic retinopathies.

Microaneurysms are seen as dot hemorrhages, which are the earliest funduscopic feature of background diabetic retinopathy. Venous beads and loops, intraretinal microvascular abnormalities (IRMAs), and soft exudates are observed in the pre-proliferative stage. Hard exudates may be found in the macular, peri-macular areas, or scattered throughout the retina. Vitreous hemorrhage, pre-retinal hemorrhage, and retinal detachment are the hallmark of the proliferative stage. Extensive blot hemorrhages herald onset of the pre-proliferative stage.

41) c.

Objective: review the trinucleotide repeats expansion and their associations.

Expanded GAA repeats of Frataxin gene on chromosome 13 are responsible for Friedreich's ataxia, while myotonia dystrophica type I results from CTG trinucleotide repeats expansion of dystrophica myotonia protein kinase (DMPK) gene on chromosome 19. Huntington's disease and Kennedy's (spinobulbar muscular atrophy) syndrome result from CAG repeats expansion, as does spinocerebellar ataxia type 1. Spinocerebellar ataxia type 8 is associated with CTG trinucleotide expansion. In fragile X chromosome syndrome locus A, CGG expansion is the culprit, while GCC expansion is seen in locus B syndrome.

42) a.

Objective: review cocaine toxicity.

Cocaine intoxication may cause intracerebral and/or subarachnoid hemorrhages (irrespective of the patient's blood pressure) as well as ischemic stroke. Agitation and psychosis are usually prominent. Hypertension may result and hypertensive crisis may ensue. Hyperthermia develop. The pupils are dilated. Myocardial ischemia and infarction may complicate the picture. Note that cocaine intoxication has 3 phases; early stimulation phase (I), advanced stimulation phase (II), and finally the co-called depressed/moribund phase (III). The symptomatology and physical signs progress from one phase to another.

43) e.

Objective: review the risk factors for cholesterol and pigment gall stones formation.

Chronic biliary infections, alcoholic liver disease, primary biliary cirrhosis, chronic hemolysis, duodenal diverticulum and truncal vagotomy are risk factors for pigment gall stones development. Cholesterol gall stones may result from obesity or rapid weight loss, estrogens, octreotide, and ileal disease or bypass.

44) b.

Objective: review lichen planus.

Lichen planus usually targets people in their 5th and 6th decades. Chronic hepatitis C infection and ingestion of beta blockers, penicillamine, quinidine, and methyldopa may result in a lichen planus-like rash but characteristically no oral involvement occurs. The oral lesions may ulcerate (usually clinically challenging). Severe scarring alopecia occurs with scalp involvement; lichen planopillaris. The skin lesions are extremely pruritic.

45) a.

Objective: review the typical clinical features of angina, and differentiate angina from other causes of chest pain.

The angina pain is typically substernal in location, which may radiate to the shoulders, neck, or jaw. Isolated shoulder pain is rare. The pain is brought about by exertion and is relieved by rest; typically, it lasts less than 5 minutes.

The pain is described as dull heaviness and may be associated with dyspnea, vomiting, and sweating. A slow and prolonged improvement should cast a doubt on the diagnosis of chronic stable angina.

46) d.

Objective: review the management of raised intracranial pressure.

Controlled hyperventilation is one of the tools that may control raised intracranial pressure (ICP). The objective is to produce $PaCO_2$ of 25 to 35 mmHg; the resulting high PaO_2 and low $PaCO_2$ would result in intracranial vasoconstriction with secondary reduction in the cerebral blood flow. Mannitol is given intravenously. Direct ventricular drainage is another way of decompressing the intracranial compartment; this is done simultaneously while measuring the intracranial pressure invasively. Pentobarbital is used for barbiturate coma. Hypotonic intravenous fluids worsen brain edema and should be avoided at all possibilities.

47) d.

Objective: review the causes and consequences of hypomagnesaemia

Hypomagnesaemia may result from pentamidine, cyclosporine, diuretics, severe diarrhea, diabetic ketoacidosis, extensive burns, famine, and malabsorption.

48) c.

Objective: have a look at bioterrorism.

This subject is becoming popular. A rapid outbreak of a zoonotic disease in a community should always make you think of bioterrorism.

49) e.

Objective: review the assessment of environmental hazards for elderly people.

You may encounter a question about geriatrics every now and then, so be well-prepared. The other factor is multiple evidence of disrepair.

50) b.

Objective: differentiate between dementia and delirium.

Rapid onset of confusion, impaired awareness, and reversed sleep-wake cycles are features of delirium. Consciousness and awareness are intact in dementia. Memory is impaired in both. Note that prominent vegetative symptoms (insomnia, fatigue, anorexia, weight loss,...etc.) favor depressive pseudo-dementia over dementia.

51) d.

Objective: review normal findings that can be seen in athletes and differentiate them from pathological changes.

Athletes have an increased cardiac output at lower heart rates; resting bradycardia is expected. Their ECG may show sinus bradycardia, first degree AV block, or Mobitz type I second degree heart block because of high vagal tone; neither Mobitz type II nor 3rd degree heart blocks are "normal,". This also applies to septal hypertrophy (? hypertrophic cardiomyopathy). Normal septal thickness is up to 11 mm; 15 mm is an asymmetric septal hypertrophy.

52) a.

Objective: review the management of Kaposi's sarcoma.

Skin involvement in Kaposi's sarcoma may take many forms, such as macules, nodules,...etc. Oral involvement usually heralds the involvement of the gut (which is usually asymptomatic); the pulmonary involvement is almost always symptomatic with dyspnea, cough,...etc. After examining the skin, one should look into the mouth, simply as part of disease's surveillance. Rushing into chemotherapy or radiotherapy would be unreasonable for the time being. The lesions may regress after starting highly active anti-retroviral therapy (HAART) with improvement in the CD4+ cell count, but not spontaneously. Human Herpes virus type 8 is implicated as an etiological factor; Castleman disease and primary effusion lymphoma have the same viral association.

53) e.

Objective: review the smoke cessation program and its benefits.

There is no nicotine inhalation as part of smoke cessation programs; this is smoking! Bupropion is contraindicated in epilepsy and eating disorders. Counseling is important as part of patient's motivation and education.

54) e.

Objective: review management of obesity and the medications used with this respect.

Sodium valproate can cause weight gain. Sibutramine and orlistat are used for weight reduction. These medications should be used in conjunction with lifestyle modification and diet recommendation, and may result in modest weight loss. Because of its noradrenergic activity, sibutramine can result in hypertension, tachycardia, and dry mouth; it should not be prescribed for patients with uncontrolled hypertension, ischemic heart disease, cardiac dysrhythmia, and stroke. It also inhibits pancreatic lipase and results in fat malabsorption, perianal soiling, and gases. Phentermine is an appetite suppressant. Topiramate is an anti-epileptic agent which induces weight loss.

55) c.

Objective: review causes of red eye.

Red eye could be acute, subacute, or chronic and the presence of ocular pain and/or diminished visual acuity requires an urgent ophthalmologic referral. The commonest cause of acute red eye is viral or bacterial conjunctivitis. Acute angle-closure glaucoma, not the chronic simple glaucoma, can result in acute red eye with headache and repeated vomiting.

56) c.

Objective: review the management of cancers of unknown primary.

Some cancer patients (3-5%) present with metastatic lesions as the sole manifestation; attempts at finding the primary site prove fruitless, even after thorough and exhausting investigations, and even post-mortally. Any malignant ascites with unknown primary in elderly females is managed as stage IV ovarian epithelial cancer. A young male presenting with a midline malignant mass with negative testicular examination and imaging should be managed as having occult extra-gonadal germ-cell tumor. A young, middle-aged, or elderly female presenting with malignant axillary lymph nodes and negative breast examination and imaging can be managed as having breast cancer.

MOCK PAPER NUMBER TWO - ANSWERS

The finding of squamous cell cancer after histopathological examination of upper cervical lymph nodes of males or females who have negative head and neck examination, should be managed according to the guidelines of squamous cell cancer of the head and neck.

57) b.

Objective: review the sites of predilection for primary spontaneous hypertensive intracerebral hemorrhage.

The history is suggestive of hypertensive intracerebral hemorrhage, which is by far the commonest cause of intracranial hemorrhage. This type of hemorrhage has the basal ganglia (putamen) as the commonest site, followed by the thalamus, lobar, cerebellar, and pontine ones in order of decreasing frequency. Although deep thalamic hematomas may decompress themselves into the 3rd ventricle, an evidence of the primary site is almost always present.

58) c.

Objective: review screening for SLE, and the antibodies that are encountered in that disease.

ANA is positive in 95% of cases of SLE patients, making it the most sensitive blood test for screening; however, it is highly non-specific. Anti-dsDNA antibodies are positive in 40-60% of "active" disease, but they are not used for screening purposes. Although anti-Sm (Smith) antibodies are highly specific for SLE, they lack sensitivity, as they are seen in only 20-30% of cases. Anti-histone antibodies are highly sensitive and specific for drug-induced lupus. Rheumatoid factor neither supports nor refutes the diagnosis of SLE, as it has nothing to do with screening or diagnosis. About 5% of SLE cases are ANA-negative throughout their course of illness; those patients usually escape detection and may be managed as adult Still's disease or some other form of vasculitis.

59) b.

Objective: review anorexia nervosa and differentiate it from bulimia.

The overall clinical picture points out towards anorexia nervosa. The clinical manifestations and the abnormal lab features result from long-term starvation as well as stress and hormonal changes which accompany it; we may encounter anemia, electrolyte imbalance, low TSH, osteopenia,...etc.

OSAMA S. M. AMIN FRCPI

Death may occur during re-feeding because of exacerbation of hypomagnesaemia, resulting in cardiac standstill. Depression frequently co-exists and should be addressed.

60) c.

Objective: review the mode of action and implementation of imatinib mesylate.

This oral medication inhibits the BCR-ABL tyrosine kinase, trying to eliminate the CML cell clone in the bone marrow. It is usually well-tolerated, but some patients are reluctant to it because of nausea and muscle cramps. It can cause transient leukopenia. This medication is being used in some gastrointestinal stromal tumors and idiopathic pulmonary fibrosis.

61) c.

Objective: review the management of aspirin poisoning.

This girl has severe poisoning as evidenced by her CNS manifestations; this requires urgent hemodialysis. Aspirin has no antidote (a distraction). The other options can be used as a bridge to hemodialysis.

62) c.

Objective: review the role of urinary sodium in differentiating pre-renal form intrinsic causes of renal failure.

Because of the resulting shock and the duration of the presentation, one can assume that she has pre-renal failure; acute tubular necrosis is too early to happen. The body tries to conserve water and sodium after activating the renin-angiotensin system, and the avid sodium retention results in low urinary sodium. In acute tubular necrosis, the tubules are no longer functional, and sodium escapes into urine (with a urinary concentration of >20 mEq/L).

63) b.

Objective: review the management of cardiac ectopics.

Benign atrial ectopic beats are very common in otherwise healthy young individuals. Unless they are frequent or troublesome, they require no treatment apart from reassurance.

64) b.

Objective: review the management of pneumothorax.

His pneumothorax was probably small to start with, but the rapid accumulation of air with time has resulted in a "tension" type of pneumothorax with hypotension and collapse. Immediate intervention is a indicated by placing a chest tube under water seal apparatus.

65) d.

Objective: review the management of LDL-cholesterol according to cardiovascular risk factors.

Coronary artery disease or its equivalents (peripheral vascular disease and diabetes) call for lowering the serum LDL cholesterol to a level blow 100 mg/dl.

66) a.

Objective: review the lab tests that are used to diagnose viral hepatitides.

Testing for antibodies against hepatitis C virus is a rapid and cheap method for diagnosing this infection; however, false-positive results do occur. The gold standard method is PCR testing, looking for the viral genome in blood. Liver biopsy assesses the disease severity and may help choose the appropriate step in the treatment. Stem "d" is used in hepatitis B infection, not hepatitis C.

67) e.

Objective: review treatment options in metastatic breast cancer.

This woman has estrogen receptor-negative metastases, but over-expression of the HER2 oncogene. Trastuzumab, a humanized monoclonal antibody, attacks this oncogene, results in tumor shrinkage, and may even prolong survival. It can be used alone as a monotherapy or combined with chemotherapy. Estrogen/progesterone receptor-positive tumors can be treated by hormonal manipulation.

68) b.

Objective: review childhood epilepsies.

This child has absence seizures. Salaam attacks are seen as episodic extensor, sometimes flexor, spasms of the trunk. The description does not fit pseudo-seizures or malingering. Complex partial seizures may have a similar presentation but do not occur with such a frequency per day; the patient does not stare all of a sudden, but continues his activity (with no recall afterwards).

69) b.

Objective: review organ involvement in CREST and compare it with systemic sclerosis.

Pulmonary complications in CREST may be in the form of recurrent aspiration (because of esophageal dysmotility). Primary pulmonary hypertension is a well-known complication, and may present as exertional breathlessness, syncope, and chest pain. Lung examination is typically normal, and right-sided failure is an advanced feature. The early subtle chest X-ray findings of pulmonary hypertension may escape detection by the inexperienced radiologist. Interstitial fibrosis is more common in the systemic diffuse variety; it is rare in CREST syndrome.

70) a.

Objective: review bullous skin diseases.

Pemphigus fits this patient's clinical scenario. Because the line of cleavage is intra-epidermal, the bullae are flaccid and rupture easily; actually, some cases may show only denuded skin areas (indicating ruptured bullae). In bullous pemphigoid, the line of cleavage is along the dermo-epidermal junction rendering the bullae tense.

71) b.

Objective: Review physical sign and their clinical significance.

Pemberton's sign is positive when a patient with multinodular goiter experiences facial congestion, breathlessness, and lightheadedness upon raising his/her arm above the head; this indicates that the goiter is compressing the thoracic outlet structures.

72) c.

Objective: review genetic causes of renal tubular defects.

Gitelman syndrome results from a mutation in the thiazide-sensitive sodium-chloride co-transporter in the distal tubule, causing a clinical picture that is indistinguishable from thiazide abuse (or therapy) and that is hypokalemia, hypochloremia, metabolic alkalosis, and low/low normal blood pressure. Bartter syndrome mutates the chloride transporter in the ascending limb of Henle's loop; the resulting frusemide abuse-like picture is similar to Gitelman syndrome, except that hypomagnesemia and hypocalciuria characterize the latter. Gitelman patients may have some skeletal manifestations.

73) d.

Objective: review the management of hereditary angioedema.

This man has been experiencing attacks of angioedema; acute attacks can be treated by fresh-frozen plasma infusion to raise the serum level of C1 esterase inhibitor. The rationale of prescribing danazole is to increase the endogenous synthesis of this enzyme. The disease is autosomal dominant.

74) b.

Objective: review the medical treatment of Crohn's disease.

Infliximab is a TNF-alpha inhibitor and is reserved for the treatment of severe fistulating Crohn's disease, rheumatoid arthritis (RA), juvenile rheumatoid arthritis, and ankylosing spondylitis. Small and minor perianal fistulae may respond to ciprofloxacin and metronidazole, and sometimes with surgical evacuation of pus. Infliximab has replaced azathioprine as a fistula healer in severe Crohn's disease. Infliximab has been associated with increased incidence of infections, particularly tuberculosis.

75) d.

Objective: review the types of assisted ventilation.

Even if you have no idea at all about the stems mentioned in the question, you must guess that CPAP is a non-invasive ventilation (via a facial or nasal masks). BiPAP can be invasive or non-invasive.

76) d.

Objective: review the side effect profile of anti-HIV medications.

All protease inhibitors, apart from atazanavir, result in abnormalities in serum lipids, in the form of elevated total and LDL cholesterol as well as triglyceride.

77) c.

Objective: review the inheritance patterns of diseases.

Properdin (a complement system regulatory protein) deficiency is an X-linked recessive disease. HLA system, blood group antigens, and α1-antitrypsin deficiency are autosomal co-dominant diseases.

78) d.

Objective: review vaccination policy with respect to influenza and pneumococcal vaccines.

Influenza vaccine is given annually, while pneumococcal vaccine is administered every 5 years if the patient has a continuous risk factor (such as COPD, alcoholism, asplenia,…etc.). Although the immunity wanes gradually after pneumococcal vaccination, a booster dose is "generally" not indicated for healthy individuals who display no risk factors for pneumococcal bacteremia.

79) d.

Objective: review the primary and secondary prophylaxis of esophageal variceal bleeding.

Propranolol is a non-selective beta blocker that is effective in primary and secondary variceal hemorrhage; atenolol is cardio-selective and has no place in these prophylaxes. Esophageal trans-section and balloon tamponade are used to arrest hemorrhage in the acute setting only. Isosorbide mononitrate is a 2nd line agent in the acute hemorrhagic setting.

80) c.

Objective: review the management of toxoplasmosis.

Approximately, 90% of toxoplasmosis patients recover from the disease without any intervention. Only 10% of cases, usually immune compromised or pregnant patients, require medical treatment. This man is healthy-looking, and needs no treatment for the time being.

81) d.

Objective: review bad prognostic signs of heart failure.

Resting tachycardia, presence and persistence of third heart sound, low blood pressure, high plasma renin activity, hyponatremia (without diuretics), low ejection fraction, high plasma adrenalin, and concomitant diastolic dysfunction are poor prognostic factors.

82) c.

Objective: review transfusion-related reactions.

The temporal profile of jaundice and the sudden fall in hemoglobin point out towards delayed hemolytic transfusion reaction (anamnestic reaction with alloantibody surge).

83) a.

Objective: review the treatment and prophylaxis of migraine.

Catamenial migraine is thought to result from low perimenstrual levels of estrogen; this is the rationale of giving low-dose estrogen during menses. Triptans are used as an abortive therapy in acute attacks. High flow, high concentration O_2 is used to terminate cluster headache attacks.

84) b.

Objective: review the etiologies of Cushing's syndrome.

Exogenous steroids use is the commonest cause of Cushing's syndrome in general, while pituitary-dependent Cushing's disease is the commonest cause in the absence of steroid administration.

85) a.

Objective: review types of maturity onset diabetes of young (MODY).

MODY type 3 is the commonest one (60-70%) and results from a mutation in the hepatic nuclear factor 1α. Glucokinase mutation causes type 2 MODY, which ranks 2nd in frequency, and producing mild hyperglycemia. Overall, these types of diabetes are usually mild and slowly progressive and can be successfully controlled with diet and oral hypoglycemic agents.

86) c.

Objective: review the classification of blood pressure.

This woman's blood pressure falls within the "high normal" blood pressure group (or pre-hypertension stage in the United States). A reading of < 120/80 mmHg is "normal". Stage I hypertension ranges between 140-159 mmHg (systolic) and 90-99 mmHg (diastolic), while stage II hypertension has values greater than 160 mmHg (systolic) or greater than 100 mmHg (diastolic).

87) c.

Objective: review the management of nephrolithiasis.

Renal transplantation corrects the current organ failure and the tubular defect that has resulted in cystinuria. The other stems are used in the management of renal stones (resulting from this tubular defect), but do not correct this patient's uremia.

88) a.

Objective: review the ANCAs.

Anti-neutrophil cytoplasmic antibodies have 2 staining patterns on indirect immunofluorescence testing; cytoplasmic (cANCA) and perinuclear (pANCA). The cytoplasmic pattern is due to interaction of the antibody with the cytoplasmic serine protease 3, while the perinuclear one indicates an interaction with myeloperoxidase. Up to 90% of Wegener's granulomatosis patients have positive cANCA, but pANCA is positive in 20% of cases only. A positive cANCA test is not diagnostic for Wegener's (except probably in isolated sinus disease or isolated glomerulonephritis), and tissue diagnosis is usually indicated; pANCA can be positive in ulcerative colitis and rheumatoid arthritis as well as other vasculitides.

89) a.

Objective: review eosinophilic pneumonia.

The clinical features and imaging are suggestive of chronic eosinophilic pneumonia. Looking for blood eosinophilia would be a reasonable step at the moment. The response to steroids is dramatic in the majority.

90) c.

Objective: review the causes of acute and chronic pancreatitis.

Alcohol consumption ranks first on the list of causes of chronic pancreatitis; gall bladder disease (stones) ranks second, followed by previous severe acute attack, hereditary pancreatitis, and autoimmune pancreatitis.

91) a.

Objective: review the management of primary hyperparathyroidism.

Surgical removal of the diseased parathyroid glands cures hypercalcemia. The other stems can be used to lower and control serum calcium. Generally, indications for surgery are: young age (<40 years), high serum calcium (>3 mmol/L), and development of complications (such as renal stones, renal impairment, severe hypertension,…etc.).

92) a.

Objective: review the diagnosis of idiopathic Parkinson's disease.

Parkinson's disease is usually idiopathic, and the constellation of asymmetric resting tremor, bradykinesia, postural instability, and the favorable response to L-dopa therapy would suffice in order to secure the diagnosis in almost all cases. The presence of any atypical feature (such as cerebellar signs, extensor planters, chorea,…etc.) should call for laboratory testing and brain imaging. Toxicology screen is reasonable when the clinical picture is suggestive, e.g., exposure to manganese, MTPT,…etc.

93) c.

Objective: review heparin-induced thrombocytopenia.

The appearance of a new thrombotic event in spite of adequate anti-coagulation should always prompt the physician search for heparin-induced thrombocytopenia; a complete blood count looking for platelets number is a simple and rapid method of screening. Serotonin release assay is the gold standard for the diagnosis. Doing ELISA for serum antiheparin-platelets factor 4 antibodies is another method. Heparin should be stopped; direct thrombin inhibitors are the anticoagulants of choice in this situation.

94) c.

Objective: review the clinical implication of BRCA1 and BRCA2 mutations.

Patients who test positive for BRCA1 and BRCA2 mutations are at risk of future development of breast and ovarian cancers. In men, prostatic cancer ranks second to breast cancer.

95) a.

Objective: review the management of various valvular heart lesions.

The patient is totally asymptomatic. Apart from regular check-up and infective endocarditis prophylaxis, he needs nothing with respect to his mild aortic stenosis for the time being.

96) b.

Objective: review the prophylaxis and management of travelers' diarrhea.

Enterotoxigenic E. coli is the cause in 50% of cases. Oral rehydration therapy is the mainstay in the treatment. Loperamide and a fluoroquinolone may have an additive value.

97) c.

Objective: review the physiological changes of pregnancy.

Many organ systems undergo certain changes during pregnancy, sometimes dramatically. The glomerular filtration rate increases by about 50% and together with increased creatinine clearance, serum creatinine becomes low. The patient's serum creatinine is considered high for this pregnant woman; renal function should be investigated thoroughly. A daily urinary loss of 150 to 300 mg of proteins is seen in normal pregnancy; higher values are abnormal.

98) d.

Objective: review risk factors for NSAID-induced peptic ulceration

Risk factors for NSAID-related gastrointestinal ulceration and hemorrhage are: advanced age; a previous gastrointestinal hemorrhagic event; higher doses of NSAIDs; the concomitant use of anticoagulants or corticosteroids; and cardiovascular diseases. Low-dose aspirin is also associated with a definite risk of peptic ulceration and upper gastrointestinal bleeding, although how this risk compares with that of other NSAIDs is not yet clear. Two main strategies have been employed to prevent the development of gastrointestinal mucosal injury in NSAID users: co-therapy with a high-dose H_2 antagonist, a PPI, or a synthetic prostaglandin analogue (misoprostol); or substitution of a traditional NSAID with a COX-2 inhibitor. Patients with no gastrointestinal risk factors should receive traditional NSAIDs. Patients with gastrointestinal risks may receive either NSAID with PPI/misoprostol or a celecoxib alone. For patients with high gastrointestinal risk factors who need NSAIDs, including those with a prior bleeding ulcer, or on concurrent anticoagulant therapy, COX-2 inhibitor plus a PPI provide the most promising gastrointestinal protection. Sodium aliginate is an antacid that is mainly used in gastroesophageal reflux disease.

99) d.

Objective: review sensitivity, specificity, and positive and negative predictive values

The so-called positive and negative predictive values of screening tests are calculated by making the following table:

	Disease present	No disease present
Positive testing	A	B
Negative testing	C	D

A positive predictive value reflects the percentage of individuals who test positive and who are actually having the disease, while the negative predictive one reflects the true percentage of individuals who are negative for the test and who don't have the disease in question.

Positive predictive value=A/(A+B); it is 85% in our case [96/(96+16)].

Negative predictive value=D/C+D; it is 95% here [84/(4+84)].

Sensitivity=A/(A+C); it is 96% in our question.

Specificity=D/(B+D); it is 84% in our case.

100) d.

Objective: review eligibility for receiving CPAP.

The patient should be reasonably conscious to cooperate with the machine; he should be able to cough and to clear and protect his airways.

Mock Paper Number Three

100 Best of Five Questions

This page was intentionally left blank

1) A 19-year old man visits the physician's office for a health checkup prior to his medical insurance. You notice craniofacial features and thumb and wrist signs of Marfan syndrome. The patient is otherwise healthy and enjoys and independent life. Which one of the following is a feature of Marfan syndrome?

    a. Hypercoagulability
    b. Risk of aortic dissection
    c. Thoracic kyphosis
    d. Increased skin elasticity
    e. Osteoporosis

2) A 16-year-old male was brought to the doctor's office because of his facial appearance. You diagnose fascioscapulohumeral dystrophy. Which one of the following is a recognized feature of facioscapulohumeral muscular dystrophy?

    a. Bilateral ptosis
    b. Wasting of both hands
    c. Tender weakness of pelvic girdle
    d. Grip myotonia
    e. Buttocks atrophy

3) A 26-year-old woman developed anemia and then was diagnosed with folate deficiency. Which one of the following statements about folate deficiency is the *correct* one?

    a. May occur in chronic alcoholics
    b. Results in microcytosis
    c. Reticulocytosis with polychromasia occur
    d. May end up with spastic paraparesis
    e. Responds to iron replacement

4) A 65-year-old man was admitted to the coronary care unit. He developed severe and prolonged substernal pain and nausea. You consider coronary insufficiency. Which one of the following is *not* a recognized ECG finding in acute myocardial ischemia?

    a. Normal ECG
    b. T-wave inversion

    c. Inversion of U-wave
    d. Short P-interval
    e. Tall R-wave in lead V_1

5) A 34-year-old woman has undergone laboratory testing for thyroid
    function. Serum TSH is 0.09 µU/ml (normal 0.5-5.0 µU/ml) and free
    T_4 is 1.2 ng/dl (0.7-1.9 ng/dl). Which one of the following does *not*
    result in low serum TSH with normal free T_4 hormone?

    a. Subclinical hypothyroidism
    b. Sick euthyroid syndrome
    c. Glucocorticoid administration
    d. Hyperthyroidism, treated with medical therapy
    e. T3 toxicosis

6) A 73-year-old man has developed progressive failing vision. You do
    fundoscopy and find maculopathy. You consider age-related macular
    degeneration. Which one of the following is *not* true with respect to
    age-related macular degeneration?

    a. A rare cause of visual loss in elderly people
    b. Family history may be positive
    c. It is of dry or wet type
    d. Can result in neovascularization
    e. Has no specific treatment

7) A 64-year-old woman presents with multiple skin bullae. Which one of
    the following does *not* result in bullous skin lesions?

    a. Lupus vulgaris
    b. Porphyria cutanea tarda
    c. Penicillamine
    d. Impetigo
    e. Diabetes mellitus

8) A 39-year-old woman developed right-sided breast mass. You find
    right-sided axillary lymph node enlargement. Histopathological
    examination of the mass reveals invasive intra-ductal carcinoma.
    Which one of the following is *not* a recognized risk factors for the
    development of breast cancer?

a. Early menarche
b. Multi-parity
c. Late menopause
d. Personal history of endometrial cancer
e. History of breast cancer in her mother

9) You test serum creatinine in a 73-year-old man who has severe global wasting. The result returns as 0.4 mg/dl. You think that this low value is attributed to the patient's age and muscle wasting. Which one of the following decreases creatinine clearance independently of the glomerular filtration rate?

a. Cimetidine
b. Ketoacids
c. Liver disease
d. Aspirin
e. Cephalothin

10) A 38-year-old woman presented with exertional breathlessness. You found atrial fibrillation. Echocardiography reveals atrial septal defect. Which one of the following is the *correct* statement about atrial septal defects?

a. Primum type is the commonest one
b. Small defects are usually asymptomatic and compatible with normal life
c. Secondum type confers a high risk of infective endocarditis
d. 12-lead ECG shows incomplete left bundle branch block
e. Chest X-ray reveals large aortic knob and arch

11) A 29-year-old man visits the Emergency Department. He has severe dyspnea. He was diagnosed with asthma before 2 years. You examine the patient and think that he has features of a life-threatening acute asthmatic attack. Features of life-threatening asthmatic attack include all of the following, *except?*

a. Bradycardia.
b. Inability to speak
c. SaO2 <90% on room air
d. Loud wheezes throughout expiration
e. PaO2 <60 mmHg

12) A 54-year-old woman was brought to the Emergency Room after ingesting 36 tablets of a medication. You resuscitate her and give her the required antidote. Which one of the following substances/medications matches its corresponding antidote?

    a.   Isoniazid-vitamin B_{12}
    b.   Benzodiazepines-nalaxone
    c.   Cyanide-fomepizole
    d.   Atropine-physostigmine
    e.   Aspirin-N acetylcysteine

13) A 34-year-old man visits the physician's office. He has developed fever, malaise, and lymphadenopathy. You take a through history and examine him properly. You consider serum sickness which was induced by a recently ingested medication. Which one of the following statements about serum sickness is the *correct* one?

    a.   Type II hypersensitivity reaction
    b.   Usually appears 7 to 10 hours after antigen exposure
    c.   There is wide-spread urticarial rash
    d.   The culprit antibody is of IgM class
    e.   Aspirin does not relieve the resulting arthralgia

14) A 19-year-old male was recently diagnosed with Gilbert's syndrome. He asked you about the disease's course and prognosis. Which one of the following is the *correct* statement with respect to Gilbert's syndrome?

    a.   Results from glucuronyl transferase over-activity
    b.   There is direct hyperbilirubinemia
    c.   Jaundice usually disappears after prolonged fasting
    d.   The prognosis is poor in the majority
    e.   Has no specific treatment

15) A young divorced female was brought by her friend to consult you. The patient said that she feels the need to check things repeatedly, perform certain routines repeatedly, and has certain thoughts repeatedly. She is unable to control either the thoughts or the activities for more than a short period of time. Which one of the following is the *correct* statement with respect to obsessive-compulsive disorder?

a. An extremely rare disease
b. Some patient may develop chronic psychosis
c. There is an overlap with nail biting
d. Patients usually uncover their symptoms and worries
e. Unemployment is protective factor against it

16) A young man was admitted to the Emergency Department because of severe non-localized abdominal pain. The patient had hallucinations, hypertension, tachycardia, and sweating. The surgeon at first thought of acute abdomen but the patient gave a history of similar attacks in the past. You found tremor and upper limbs weakness. His abdominal ultrasound in unremarkable. Which one of the following statements about acute intermittent porphyria is the *correct* one?

a. Caused by partial deficiency of porphobilinogen deaminase
b. The resulting abdominal pain is ischemic in origin
c. Hypernatremia is common
d. Freshly voided urine looks grossly abnormal and very dark
e. Acute attacks can be treated with intravenous ringer solution

17) A 23-year-old man visits the physician's office for a scheduled follow-up. He was diagnosed with type I diabetes before 3 months. His home readings of fasting and random blood sugars are very high. You do HbA1c assessment and this turns out to be "5.3". All of the following may falsely lower HbA1c value, *except?*

a. Beta thalassemia major
b. Paracetamol therapy
c. Chronic renal failure
d. Hemolysis
e. Menorrhagia

18) A young man was recently diagnosed with celiac disease because of steatorrhea and weight loss. Which one of the following about celiac disease is *true?*

a. Common in Africans and Asians
b. Associated with HLA Cw6
c. The severity depends on the thickness of bowel wall involvement
d. Serum IgA anti-gliadin antibody is not useful to monitor treatment compliance

e. Can be complicated by T-cell lymphoma

19) A 39-year-old homosexual man was recently diagnosed with AIDS because he had developed an AIDS-defining illness. Which one of the following AIDS-defining illnesses needs a lab evidence of HIV infection?

a. Tracheal candidiasis
b. Kaposi sarcoma in patients < 60 years
c. Recurrent non-typhoidal Salmonella bacteremia
d. Primary CNS lymphoma in patients <60 years
e. Progressive multifocal leukoencephalopathy

20) A 54-year-old man did some blood tests. His total serum cholesterol was 110 mg/dl. Which one of the following may result in reduction of serum total cholesterol?

a. Malignancy
b. β-blockers
c. Nephrotic syndrome
d. Chronic cholestasis
e. Alcohol abuse

21) A 65-year-old man developed ST-elevation myocardial infarction. Which one of the following is *not* to prevent future ischemic events?

a. Simvastatin
b. Lisinopril
c. Metoprolol
d. Aspirin
e. Sublingual nitrate

22) A patient with diagnosed with Buerger's disease. His left distal lower limb will be amputated because of critical ischemia. Which one of the following is the *correct* statement about Buerger's disease?

a. Affects arteries only
b. Mainly targets elderly females
c. Proximal and distal pulses are feeble or absent
d. Always affects one limb
e. Skin ulcers are observed at nail margins

23) You are examining the chest X-ray of this 39-year-old man who was diagnosed with allergic bronchopulmonary aspergillosis (ABPA) yesterday.

   a. There is peripheral blood eosinopenia
   b. Fleeting pulmonary infiltrates are common
   c. Peripheral bronchiectasis occurs
   d. Negative skin response to Aspergillus antigen is expected
   e. Serum precipitin testing against Aspergillus is usually strongly positive

24) A 23-year-old male developed cardiomyopathy secondary to doxorubicin therapy for acute lymphoblastic leukemia. Which one of the following is *not* toxic to the myocardium?

   a. Candesartan
   b. Cocaine
   c. Methotrexate
   d. Sodium stibogluconate
   e. Lithium carbonate

25) A 43-year-old female presents with stridor and carpo-pedal spasms. She improves dramatically when intravenous calcium gluconate was administered. Which one of the following does *not* result in hypocalcemia?

   a. Alcoholism
   b. Milk-alkali syndrome
   c. Hyperphosphatemia
   d. Short bowel syndrome
   e. Nephrotic syndrome

26) A 61-year-old man was found to have primary proliferative polycythemia after doing complete blood counts as part of his general medical check-up. Which of the following is the *correct* statement about primary proliferative polycythemia?

   a. There is a reduced red cell mass
   b. The ESR is elevated
   c. Serum vitamin B_{12} is usually elevated
   d. Absence of splenomegaly is usual

e. Urinary erythropoietin level is very high

27) A 36-year-old man visits the Emergency Room because of sever left-sided loin pain and hematuria. Abdominal ultrasound reveals a stone in the left kidney pelvis. Which one of the following is *not* a recognized risk factor for renal stone formation?

a. Hypercitraturia
b. Hyperoxaluria
c. Hypercalciuria
d. Hyperuricosuria
e. Urinary tract obstruction

28) Because of his gross skeletal abnormalities, a 46-year-old man was diagnosed with acromegaly. Which one of the following is *incorrect* with respect to acromegaly?

a. Most patients demonstrate macroglossia
b. Decreased sweating is very common
c. Excessive sebum secretion occurs
d. Kyphosis may develop
e. Patients may develop hypertension

29) Because of excessive fatigue, sleep disturbances, and wide-spread pain and allodynia, a 31-year-old woman was diagnosed with fibromyalgia. Which one of the following is *true* about fibromyalgia?

a. The ESR is raised
b. There is neutrophilic leukocytosis
c. Peripheral joints are warm and red
d. Headache should be absent
e. May respond to tricyclic antidepressants

30) A 31-year-old woman presents with bilateral failing vision. You find bilateral secondary optic atrophy. You do brain imaging and lumber puncture. Your diagnosis is idiopathic pseudotumor cerebri. With respect to idiopathic pseudotumor cerebri, which one of the following is the *correct* statement?

a. The intracranial pressure is reduced
b. Horizontal diplopia results from abducens nerve palsy

   c.   Low-grade fever is usually occur

   d.   There is leucopenia

   e.   Pancephalic headache is usually absent at the time of diagnosis

31) A 54-year-old woman was prescribed tamoxifen after undergoing breast lumpectomy. The lump turned out to be malignant. Which one of the following is not *true* about tamoxifen?

   a.   Displays both anti-estrogen and estrogen agonistic activities

   b.   Useful in treating early and advanced breast cancer

   c.   Reduces the risk of endometrial cancer

   d.   Reduces the risk of developing cancer in the other breast

   e.   Has a favorable effect on serum lipid profile

32) A 24-year-old man feels sad, anxious, empty, hopeless, helpless, worthless, guilty, irritable, angry, ashamed most of the time. Which one of the following is *not* a feature of major depression?

   a.   Weight loss

   b.   Early morning insomnia

   c.   Poor concentration

   d.   Psychosis

   e.   High risk of homicide

33) Because of chronic bloody diarrhea, a 29-year-old man was diagnosed with ulcerative colitis. All of the following are correct with respect to ulcerative colitis, *except*?

   a.   Usually develops in non-smokers

   b.   The rectum is involved lately in the course

   c.   The terminal ileum may be involved

   d.   Appendectomy before the age of 20 years is protective against it

   e.   Disease severity is lower in active smokers

34) A young man has developed skin lesions, which turn out to be tinea versicolor. Which one of the following is not true about tinea versicolor?

   a.   There are fine skin scales

   b.   White macules develops in the affected skin area

   c.   The lower part of the body is mainly involved

    d.   Most lesions itch

    e.   A differential diagnosis of vitiligo

35) A young man was diagnosed with selective IgA deficiency and asked you about the disease and its complications. Which one of the following statements about selective IgA deficiency is *true*?

    a.   Serum levels of IgM antibodies are also low

    b.   Atopy is rarely associated

    c.   Selective IgG subclass 2 deficiency may coexist

    d.   It is protective against autoimmune diseases

    e.   Blood transfusion is the usual treatment

36) A young female was diagnosed iron deficiency anemia and she ingested daily iron tablets. Her pharmacist told her that iron tablets should not be ingested with milk because iron will be less absorbed. Which one of the following statements about drug-drug interaction is *correct*?

    a.   Aspirin increases serum ascorbic acid

    b.   Salphasalazine impairs folate absorption

    c. H_2-blockers enhance iron absorption

    d.   Phenytoin decreases vitamin K turnover

    e.   Diuretics decrease urinary excretion of vitamin B_6

37) A 26-year-old woman got pregnant. This was her first pregnancy. She visited you to have a general medical check-up. She was in her 3rd trimester. You examined her and did some tests. Which one of the following is *not* an expected cardiovascular physiological change during pregnancy?

    a.   The stroke volume rises

    b.   The diastolic blood pressure remains unchanged or rises slightly

    c.   The resting heart rate falls

    d.   A 3rd heart sound is normally heard

    e.   Inferior Q wave appears on 12-lead ECG

38) A 65-year-old man visits his physician. He has developed failing vision. The funduscopic examination is consistent with central retinal artery occlusion. With respect to central retinal artery occlusion, choose the *correct* statement?

a. Produces gradual visual loss
b. Fundoscopy reveals hyperemic retina
c. Cherry-red macular spot is observed after recovery
d. Optic disc atrophy is an early finding
e. The prognosis is poor in most patients

39) A 34-year-old man complains of excessive day-time sleepiness. You assess him and diagnose obstructive sleep apnea. Which one of the following is *not* a risk factor for obstructive sleep apnea?

a. Hyperthyroidism
b. Obesity
c. Receding chin
d. Acromegaly
e. Large tonsils

40) A 56-year-old man complained of malaise, fever, abdominal discomfort, and weight loss. Eventually, he was diagnosed with hairy cell leukemia. With respect to this type of leukemia, which one is the *correct* statement?

a. It is a malignancy of T-lymphocyte
b. There is pancytopenia
c. Massive splenomegaly is usually detected
d. Monocytopenia is characteristic
e. The bone marrow can be easily aspirated

41) A 65-year-old smoker man develops persistent dry cough. He admits to losing weight and feels tired all the time. You order chest X-ray. Your preliminary suspicion is lung cancer. Plain chest X-ray films in lung cancer may display all of the following, *except?*

a. Normal film
b. Lung collapse
c. Elevation of a hemi-diaphragm
d. Bilaterally small hilar areas
e. Persistent infiltrates

42) A 43-year-old obese woman developed anasarca. Urine examination uncovers nephrotic-range proteinuria. Renal biopsy reveals focal segmental glomerulosclerosis. Which one of the following is the *incorrect* statement with respect to focal segmental glomerulosclerosis?

    a.   Can be secondary to heroin abuse
    b.   Impaired renal function is usually seen at the time of diagnosis
    c.   Eventually leads to end-stage renal disease
    d.   The proteinuria is non-selective
    e.   Responds favorably to steroids

43) A 22-year-old woman was admitted to the Emergency Room because of toxic shock syndrome. With respect to toxic shock syndrome, which one of the following is the **correct** statement?

    a.   Caused by certain strains of Staphylococcus albus
    b.   It is an afebrile illness
    c.   Blood culture is negative for the offending bacteria
    d.   Renal failure should not develop
    e.   Skin desquamation is uncommon

44) A patient was admitted to the hospital after developing Kawasaki's disease. Which one of the following statements about this disease is *correct?*

    a.   Mainly develops in middle-aged males
    b.   The lip swells and many bullae form
    c.   Non-suppurative cervical lymphadenopathy occurs in many patients
    d.   Coronary artery spasm may develop
    e.   Does not respond aspirin

45) A 43-year-old man underwent sophisticated investigations to find out the nature of a suprarenal mass. The diagnosis turned out to be pheochromocytoma. Choose the *wrong* statement with respect to pheochromocytoma?

    a.   Hypertension may be sustained rather than paroxysmal
    b.   Postural hypotension may be detected
    c.   Serum TSH is normal
    d.   Most patients gain weight

e. Recurrent abdominal pain occurs

46) A 45-year-old man develops severe upper abdominal pain and vomiting. He is alcoholic. Serum amylase and lipase are elevated. Which one of the following is *not* a complication of acute pancreatitis?

a. Adult respiratory distress syndrome
b. Renal failure
c. GIT bleeding
d. Splenic vein thrombosis
e. Hypoglycemia

47) A 67-year-old man was admitted to the Emergency Room because of acute digoxin poisoning. He had congestive heart failure and permanent atrial fibrillation. Which one of the following is *not* a consequence of this acute toxicity?

a. Vomiting
b. Xanthopsia
c. Hypokalemia
d. Bradycardia
e. Ventricular tachycardia

48) A 15-year-old boy is brought to you by his parents. You notice cluttered speech and stereotypic hand-flapping movements, atypical social development, shyness, limited eye contact, memory problems, and difficulty with face encoding. Initially you think of autism. However, you examine him further and do some tests. Your final diagnosis is fragile X mental retardation. Which one of the following is consistent with the diagnosis of fragile X-mental retardation?

a. Macroorchidism
b. Receded chin
c. Small rudimentary ears
d. Aortic stenosis
e. Low pitched voice

49) A 23-year-old man visits you because of exertional breathlessness. You do chest X-ray, 12-lead ECG, and transthoracic echocardiography. You find asymmetric septal hypertrophy. Regarding the treatment of hypertrophic cardiomyopathy, which one of the following statements is the *correct* one?

a. Digoxin is helpful
b. Hydralazine is used to lower the blood pressure and reduce dyspnea
c. Beta blockers are first-line agents
d. Calcium channel blockers are contraindicated
e. Diuretics should always be added to the treatment plan

50) A 58-year-old man was diagnosed with limited stage small-cell lung cancer. He asks if radiotherapy, with or without chemotherapy is curable. Your answer was no. Radiotherapy, as a monotherapy, *cannot* cure which one of the following malignancies?

a. Laryngeal squamous cell cancer
b. Breast adenocarcinoma
c. Hodgkin's disease
d. Cervical cancer
e. Prostatic adenocarcinoma

51) A 67-year-old man complains of dry cough and breathlessness on exertion. He has permanent atrial fibrillation and hypertension. He takes many daily medications. The patient's DLCO is reduced. What medication is responsible for the man's current complaint?

a. Bisoprolol
b. Digoxin
c. Amiodarone
d. Warfarin
e. Diltiazem

52) A 72-year-old woman presents with a 1-year history of progressive bilateral hand resting tremor, rigidity, and hypokinesia. You find bilateral extensor planter reflexes. What would you do *next*?

a. Brain MRI
b. Start L-dopa preparation

c. Give benztropine

d. Prescribe low-dose amantadine

e. Observe

53) A 13-year-old boy has bronchiectasis, steatorrhea, global wasting, and nasal polypi. Sweat test is positive. His cousin died of the same disease. Which chromosome carries the mutated gene responsible for this illness?

a. 1

b. 11

c. 7

d. 17

e. X

54) A young male presents with bilateral facial palsy, right knee swelling and pain, and dropped beats. He has just returned from a long trip in Europe. What is the characteristic initial skin manifestation of this disease?

a. Erythema nodosum

b. Erythema annulare

c. Erythema gyratum repens

d. Erythema chronica migrans

e. Erythroderma

55) A 31-year-old HIV-positive man presents with exertional dyspnea, dry cough, fatigue, low grade fever, and peri-hilar pulmonary infiltrates. His DLCO is reduced and there is a low SaO_2. *Pneuomocystis carinii* pneumonia usually develops when the CD4+ cell count falls below?

a 500/mm³

b. 400/mm³

c. 300/mm³

d. 200/mm³

e. 100/mm³

56) A 49-year-old female underwent bronchoscopic examination because of chronic cough. The broncho-alveolar lavage (BAL) fluid turned out to be normal. What is the commonest cell type observed in normal BAL cytological examination?

a. Neutrophil
b. Alveolar macrophage
c. Eosinophil
d. Basophil
e. Small lymphocytes

57) A 30-year-old black male complains of dry irritative cough for 3 weeks. You find bilateral hilar lymph node enlargement as well as bilateral interstitial pulmonary parenchymal infiltrates on plain chest X-ray films. He has purplish plaques on his nose and ear lobules. ECG reveals 2nd degree AV block, Mobitz type II. He denies any lapses in consciousness or pre-syncope. How would you treat him?

a. Permanent pacemaker implantation
b. Bone marrow transplantation
c. Carvedilol
d. Prednisolone
e. Highly active anti-retroviral therapy

58) A 32-year-old man, who has had recurrent urinary tract infections and obstruction over the past 3 years, presents with loin pain. Plain abdominal film shows multiple renal stones. What kind of stones she has?

a. Cysteine
b. Xanthine
c. Calcium phosphate
d. Magnesium ammonium phosphate
e. Calcium oxalate

59) A 54-year-old man presents with pallor, jaundice, and macrocytosis for several months. His feet numb, and his personality has changed. He underwent abdominal surgery 5 years ago because of a suspicious-looking mass. What kind of operation do you think he underwent?

a. Pancreatic resection
b. Colonic resection
c. Uretro-sigmoidostomy
d. Anastomosis between common bile duct and duodenum
e. Ileal resection

60) A 17-year-old high school girl becomes suspicious about everything. Her roommate says that the patient is not participating in any school activity, and she thinks the manager will poison her because she is cleaver. What would you do *next*?

    a. Give amitriptyline
    b. Electroconvulsive therapy
    c. Echocardiography
    d. Brain CT scan
    e. Blood and toxicology screen

61) A 27-year-old woman, who has been given a diagnosis of systemic lupus erythematosus (SLE) 1 year ago, presents with a flare-up of her disease. Which one of the following is a lab marker of disease activity in SLE?

    a. Serum LDH
    b. Antinuclear factor
    c. Plasma C3 level
    d. aPTT
    e. Neutrophil count

62) A 32-year-old male complains of bifrontal headache for several months. Brain MRI reveals a 20 x 21 x 17 mm mass within the sella turcica. Serum prolactin is 1200 u/L. The cause of this hyperprolactinemia is?

    a. Pituitary prolactinoma
    b. Disconnection hyperprolactinemia
    c. Adjacent hypothalamic tumor
    d. Lung cancer secondary tumor
    e. Stress

63) A 44-year-old man is brought to the Acute and Emergency Department after developing of hematemesis. His blood pressure is 80/40 mm Hg. The most important initial step in the management is?

    a. Volume resuscitation
    b. Esophagoscopy
    c. Sigmoidoscopy
    d. Platelets counts

e.   Blood coagulation studies

64) A 14-year-old girl presents with repeated seizures. She has a round
   face and a short stature. Serum calcium is 6.5 mg/dl and serum
   parathyroid hormone is raised. Serum phosphate is 5.7 mg/dl. Renal
   function is normal. What would you do *next*?

   a.   X-ray of the hands
   b.   Ultrasonography of the parathyroid glands
   c.   Repeat blood urea and serum creatinine
   d.   MRI of the pituitary
   e.   Serum sodium

65) A 32-year-old man presents with an episode of supraventricular
   tachycardia, which has resulted in palpitation. His blood pressure is
   120/70 mmHg during the attack. He is otherwise healthy, takes no
   medications, and he denies doing drugs. How would you treat?

   a.   Intravenous propranolol
   b.   Amiodarone digoxin
   c.   Synchronized DC shock
   d.   Carotid massage
   e.   Intravenous digoxin

66) A young man has recurrent attacks of severe periorbital headache
   associated with partial drooping of the eyelid and excessive tearing.
   Eye examination is normal. These attacks have been occurring at night
   since 2 weeks. He reports a similar scenario almost 1 year ago which
   lasted for 1 month. What is the likely diagnosis?

   a.   Migraine with aura
   b.   Tension headache
   c.   Intermittent angle-closure glaucoma
   d.   Cluster headache
   e.   Anterior uveitis

67) A 64-year-old man complains of lassitude, abdominal discomfort,
   constipation, and unintentional weight loss for 5 months.
   Sigmoidoscopy is normal. On doing barium enema, you notice apple-
   core sign, high up in the left descending colon. What does the patient
   have?

a. Chronic ischemic colitis
b. Colonic cancer
c. Angiodysplasia of the colon
d. Diverticular disease of the colon
e. Crohn's disease

68) A 16-year-old high school girl presents with global wasting, bradycardia, and cold extremities. She denies being on a diet. Her mother says that the patient eats almost nothing every day. You think of anorexia nervosa. Which one of the following would be inconsistent with your provisional diagnosis?

a. Anemia
b. Hypokalemia
c. Raised serum cholesterol
d. Normal menstrual cycle
e. Raised serum growth hormone

69) A 58-year-old man presents with bilateral pitting leg edema and periorbital puffiness. He is a life-long smoker and his cough has worsened recently. He says that his urine is foamy. Urine examination reports a bland sediment. Which one of the following explains his new presentation?

a. Liver cirrhosis
b. Membranous nephropathy
c. Bladder outlet obstruction
d. Wegener's granulomatosis
e. Chronic renal failure

70) A 25-year-old patient develops seizures 2 days after a severe bout of gastroenteritis. He has not discontinued his medication, which were prescribed for his psychiatric disease. Which medication would be responsible for these seizures?

a. Haloperidol
b. Olanzapine
c. Nortriptyline
d. Fluoxetine
e. Lithium

71) A 39-year-old alcoholic male has progressive abdominal distension. Examination reveals a tinge of jaundice, clubbing, and signs of fluid in the peritoneum. Today, he has developed fever and delirium. The peritoneal fluid contains 1000 neutrophils/mm³. What is the cause of his recent presentation?

    a.   Hepatic encephalopathy
    b.   Hepatocellular carcinoma
    c.   Spontaneous bacterial peritonitis
    d.   Chronic hepatitis C infection
    e.   Secondary peritoneal carcinomatosis

72) A 60-year-old woman visits the physician's office for her annual check-up. Examination reveals a clinically euthyroid multinodular goiter, which has not changed in size since the last visit 1 year ago. Serum TSH is 0.03 mU/L and serum free T4 is 29.7 pmol/L (normal 10.3- 30.6). This woman is at risk of developing which one of the following?

    a.   Myocardial infarction
    b.   Dementia
    c.   Peripheral neuropathy
    d.   Osteoporosis
    e.   Acute myeloid leukemia

73) A 13-year-old male develops gum bleeding, nosebleed, and lower limb ecchymosis over 2 days. Apart from these bleeding sites, the examination is unremarkable. His complete blood picture shows a hemoglobin of 14 gm/dl, total white cell count of 5400 cells/ml³, platelets count of 7000/ml³, and no immature cells. What is the cause of this boy's presentation?

    a.   Thrombotic thrombocytopenic purpura
    b.   Idiopathic thrombocytopenic purpura
    c.   Aplastic anemia
    d.   Henock-Shönlein purpura
    e.   Acute lymphoblastic leukemia

74) A patient is referred to the chest clinic for further assessment. He has exertional dyspnea and dry cough. You find bilateral upper lung zones fibrosis on plain chest X-ray films. All of the following are causes of predominantly upper lung zones fibrosis, *except?*

   a. Asbestosis
   b. Silicosis
   c. Ankylosing spondylitis
   d. Cystic fibrosis
   e. Chronic farmer's lung

75) A 47-year-old man presents with a flare-up of his chronic plaque psoriasis. Which one of the following is **not** a trigger for this flare?

   a. Sun exposure
   b. Atenolol
   c. Lithium
   d. Streptococcal throat infection
   e. Trauma

76) A 23-year-old woman presents with recurrent chest infections and steatorrhea. Both of them are well-controlled with multiple medications. She has palpable lymph nodes and a palpable spleen. Her final diagnosis is common variable immune deficiency syndrome. What is the risk of passing this disease to her children?

   a. 100%
   b. 50%
   c. 25%
   d. 1%
   e. Zero percentage

77) A 40-year-old man presents with impotence and arthralgia. He reports dropped beats. His glucose tolerance test is impaired. Serum transaminases are raised and you detect very high serum ferritin. Which HLA haplotype is associated with this disease?

   a. HLA DR3
   b. HLA B5
   c. HLA B27
   d. HLA A3

e.   HLA DR4

78) You conduct a research about a novel medication called Medisan. This drug acts on the brain to prevent seizure occurrence after head trauma. You find that 10% of patients who ingested the medication had not developed seizures while 6% of patients who did not take Medisan had not developed seizures. How many patients do you need to treat to prevent seizure occurrence after traumatic brain injury using Medicsan?

    a.   100
    b.   50
    c.   25
    d.   10
    e.   4

79) A 54-year-old female presents with sudden severe occipital headache after which she develops a short lapse in consciousness. Brain CT scan shows massive subarachnoid hemorrhage. Which medication you give to reduce the incidence of cerebral vasospasm?

    a.   Nifedipine
    b.   Hydralazine
    c.   Esmolol
    d.   Nimodipine
    e.   Papaverine

80) A 29-year-old male receives treatment for stage II testicular seminoma. His pretreatment serum alpha fetoprotein (AFP) is raised. What is the cause behind this abnormal lab finding?

    a.   The tumor secretes AFP
    b.   It is probably not a pure seminoma
    c.   A technical error
    d.   Co-existent liver disease
    e.   It is a normal phenomenon

81) A46-year-old patient is referred urgently to the transplant unit because of fulminant hepatic failure. Which one of the following is the commonest cause of acute liver failure in the Western World?

    a.   Hepatitis A infection
    b.   Mushroom poisoning
    c.   Paracetamol poisoning
    d.   Carbon tetrachloride poisoning
    e.   Hepatitis B infection

82) A 65-year-old male presents with recurrent gouty arthritis. He has multiple diseases, for which he takes many daily medications. Which one of the following medications does *not* increase the urinary excretion of uric acid?

    a.   Probenecid
    b.   High-dose aspirin
    c.   Losartan
    d.   Cyclosporine
    e.   Interleukin-6

83) A 54- year-old woman presents with a slowly enlarging dome-shaped nodular lesion on her nose. The nodule has a pearly translucent surface with telangiectasia. What does he have?

    a.   Keratoacanthoma
    b.   Basal cell carcinoma
    c.   Malignant melanoma
    d.   Junctional nevus
    e.   Metastatic deposit

84) A 32-year-old man presents with recurrent nosebleeds. He is otherwise healthy and denies doing drugs. Which one of the following is *not* a common cause of epistaxis?

    a.   Idiopathic
    b.   Foreign body
    c.   Osler-Weber-Rendu disease
    d.   Viral rhinitis
    e.   Aspirin therapy

85) A 65-year-old female presents with fatigue, constipation, body aches and pain, and pins and needles sensation in her hands. Her skin is dry and thick. Which one of the following would you choose initially to confirm or refute your provisional diagnosis?

a. Serum free T_3
b. Serum TSH
c. Serum TRH
d. Serum prolactin
e. Thyroid radioisotope scanning

86) A 14-year-old female presents with a papaulo-vesicular skin rash over her ankles and cubital fossae. The rash is itchy and the involved skin appears thick. Her aunt and grandmother had the same illness when they were young. The first-line agent in the treatment of this skin disorder is?

a. Topical triamcinolone
b. Oral cyclosporine
c. Topical emollients
d. PUVA
e. Topical vitamin D

87) A 54-year-old woman presents with difficulty rising from a low chair. She feels breathless on exertion and there are dropped beats. Her pelvic girdle is tender. Serum anti-Jo1 antibody is positive. Skin examination is unremarkable. Most likely, her diagnosis is?

a. Dermatomyositis
b. Osteomalacia
c. Statin-induced myositis
d. Adult polymyositis
e. Muscle trichinosis

88) A 63-year-old male presents with 3 attacks of generalized seizures over a matter of 1 week. Otherwise, he is healthy. He denies history of head trauma. Ingestion of alcohol or illicit drugs were absent in the history. All of the following steps are appropriate in the management of this patient, *except*?

a. Start sodium valproate
b. Order brain CT scan with contrast
c. Arrange for echocardiography
d. Arrange for EEG
e. Do blood and urinary toxicology screen

89) A 43-year-old female presents with poor exercise tolerance. She admits to having heavy menses. Apart from pallor of skin and mucosal surfaces as well as a cardiac flow murmur, her clinical examination is unremarkable. Her investigations most likely reveal all of the following, *except*?

    a.   Microcytosis
    b.   Elevated platelets count
    c.   Low serum transferrin saturation
    d.   Low serum iron binding capacity
    e.   Normal white cell count

90) A 43-year-old man presents with fatigue, jaundice, and hematemesis. Further work-up uncovers the presence of liver cirrhosis. Serum HBs antigen and IgG anti-HBc antibodies are positive. Which one of the following statements regarding this virus is *true*?

a.   It is an RNA virus
b.   Usually transmitted via contaminated food or drink
c.   Never results in hepatocellular carcinoma
d.   Chronic infections can be treated with adefovir
e.   The precore mutant forms of the virus respond favorably to treatment

91) A 63-year-old male is brought to the Emergency Room. He has been experiencing acute severe piercing central chest pain, radiating to the mid-back over the past 1 hour. He has long standing hypertension and ischemic heart disease. His current 12-lead ECG is similar to the one that was done 1 month ago. Chest X-ray is normal. What do you think he has developed?

    a.   Osteoporotic dorsal spine fracture
    b.   Aortic dissection
    c.   Acute pericarditis
    d.   Non-ST segment elevation myocardial infarction
    e.   Diffuse esophageal spasm

92) A 2-year-old male is brought by his parents to the doctor's office with recurrent otitis media and sinusitis. No local lesion has been found. Investigations reveal serum IgG of 50 mg/dl and undetectable levels of IgM, IgA, and IgE. His maternal uncle had the same problem. The immunological defect of this disease lies in?

a. Pre-B cells
b. T-lymphocytes
c. Plasma cells
d. Pleuripotent stem cells
e. Neutrophils

93) A 32-year-old male presents with bilateral non-pitting edema. The edema can be detected up to the mid calves. It has been progressing over many years and the skin of the leg is elephant-like in texture. All of the following are true regarding this condition, *except*?

a. It can be sporadic
b. Diuretics are useful in alleviating the edema
c. The edema is pitting at the beginning of the illness
d. The patient is at risk of cellulitis
e. Radionuclide lymphoscintigraphy is diagnostic

94) A 19-year old male presents with a 3-day history of fever, dry cough, fatigue, and mild shortness of breath. His chest X-ray shows multifocal infiltrates and bilateral hilar lymphadenopathy. Serum cold-agglutinin is positive. What microorganism is responsible for this man's illness?

a. L. pneumophila
b. Chlamydia trachomatis
c. E.coli
d. Mycoplasma pneumoniae
e. Actinomycetes

95) A middle-aged patient presents with poor exercise tolerance, palpitations, and orthopnea. Examination and echocardiography are suggestive of severe mitral regurgitation. Which one of the following indicates that the regurgitation is severe?

a. Site of the murmur
b. The presence of systolic click
c. Third heart sound
d. Fever
e. Concomitant aortic regurgitation

96) A 66-year-old man was brought to the physicians' office by his family.
He had dementia, unstable gait, and urinary incontinence. Brain CT
scan showed communicating hydrocephalus and a diagnosis of normal
pressure hydrocephalus is being considered. The lateral ventricles
communicate inferiorly with which one of the following?

   a.   Subarachnoid space
   b.   Third ventricle
   c.   Forth ventricle
   d.   Cerebral venous sinuses
   e.   Cerebral aqueduct

97) A 32-year-old man undergoes skin testing with purified protein
derivative. After 72 hours, his tuberculin test reveals an induration of 8
mm, which was considered positive by the infectious diseases
physician. This patient is?

   a.   Healthy
   b.   Intravenous drug abuser
   c.   A recent immigrant from a highly endemic country
   d.   HIV infected
   e.   A health care worker

98) A 50-year-old female presents with recurrent hematuria. Ultrasound of
the abdomen shows polycystic kidneys. All of the following are seen in
this disease, *except*?

   a.   Aortic regurgitation
   b.   Abdominal aortic aneurysm
   c.   Hepatic cysts
   d.   Intracranial Berry aneurysm
   e.   Hypermobile joints

99) A 43-year-old patient is admitted to the intensive care unit. He has
mixed metabolic acidosis and respiratory alkalosis. The patient has
developed severe aspirin poisoning. Which one of the following is the
commonest cause of this combined acid/base disturbance?

   a.   Gram-negative sepsis
   b.   Ethylene glycol poisoning
   c.   Ammonium chloride ingestion

    d.   Acute renal failure

    e.   Renal tubular acidosis

100)   A 10-year-old HIV-positive child will be vaccinated in the foreseeable future. Which one of the following vaccines is *contraindicated* in HIV-infected individuals?

    a.   Measles

    b.   Mumps

    c.   BCG

    d.   Killed polio

    e.   DTP

Mock Paper Number Three

Answers

This page was intentionally left blank

1) b.

Objective: differentiate between Marfan syndrome and homocystinuria.

Osteoporosis and hypercoagulability are features of homocystinuria. In Marfan syndrome, aortic root diameter >40 mm requires surgical repair, especially prior to pregnancy. Premature osteoarthritis is common because of joint hypermobility. Increased skin elasticity is not seen in both conditions. However, Marfan patients may demonstrate unexplained skin stretch marks. Spinal scoliosis and thoracic lordosis, pectus excavatum, and pectus carinatum may develop. In spite of joint hypermobility, some patients may demonstrate limited mobility of the hip joints, due to the femoral heads protruding into abnormally deep sockets.

2) a.

Objective: review facioscapulohumeral muscular dystrophy.

Bilateral ptosis occurs. Diplopia does not develop, because the extra-ocular muscles are not involved. The hand muscles are characteristically spared. Tender/weak pelvic girdle refers to polymyositis. Grip myotonia is a feature of myotonia dystrophica. Tibialis anterior is the only muscle outside the facial and shoulder areas to be involved; therefore, foot drop may ensue.

3) a.

Objective: differentiate between folate and vitamin B_{12} deficiencies.

Folic acid deficiency is common in alcoholics, chronic hemolytic states, and malabsorption. Macrocytosis is observed on peripheral blood films. Folate deficiency is one of the causes of anemia with *low* reticulocyte count (a differential diagnosis of aplastic anemia, but the former displays a hypercellular marrow while the marrow in aplastic anemia is hypocelluar or acellular). Subacute combined degeneration of the cord and cerebral dysfunction (dementia, personality changes,…etc.) are seen in vitamin B_{12} deficiency, not folate deficiency. The anemia of vitamin B_{12} deficiency may respond partially to folate, but this would exacerbate the neurological dysfunction; therefore, folate should not be given to patients with "unexplained" anemia. Iron deficiency is a common accompaniment of vitamin B_{12} and folate deficiencies and should always be looked for and corrected.

4) d.

Objective: review the ECG findings of myocardial ischemia.

The 12-lead ECG could be entirely unremarkable in acute ischemia, especially early in the course. The T-wave inverts because of abnormalities in ventricular repolarization. An inverted U-wave reflects severe myocardial ischemia. A short PR-interval is seen in Wolf-Parkinson-White syndrome and tachycardia states. A new appearance of tall R wave in patients with ischemic chest pain indicates infarction of the posterior wall of the heart.

5) a.

Objective: review the causes of suppressed serum TSH with normal free T_4.

The combination of low serum TSH *and* normal serum free T_4 hormone is encountered in subclinical hyperthyroidism (free T_4 is high normal), sick euthyroid syndrome, use of glucocorticoids or octreotide, and isolated T_3 toxicosis (the serum TSH is suppressed with raised serum T_3 and normal serum free T_4). In hyperthyroid patients who have received medical therapy, the TSH usually lags (for some time) behind the corrected free T_4 level; during this period, testing the thyroid function may reveal low serum TSH with normal serum free T_4 hormone. This also applies to viral/subacute thyroiditis, on their way of recovery.

6) a.

Objective: review the 2 types of age-related macular degeneration.

Age-related macular degeneration is the leading cause of permanent visual loss in elderly people; it is more common in white females with a history of smoking. It may be of dry or wet type. Choroidal neovascularization is seen in the wet type; hemorrhage from these abnormal vessels can cause sudden visual loss. No specific treatment is present; however, some patients may benefit from low vision aids.

7) a.

Objective: review causes of bullous skin lesions.

Bullous skin eruption with target and iris-like lesions is observed in erythema multiforme. Milia, skin scars, photosensitivity, hyperpigmentation, and hypertrichosis are the skin manifestations of porphyria cutanea tarda. Penicillamine may result in pemphigus. A honey-colored scab is usually seen in impetigo. Bullae may form in the skin of the feet of diabetic patients; so-called bullous diabeticorum. Lupus vulgaris results in painful skin nodules.

8) b.

Objective: review risk factors for breast cancer.

The following are risk factors for breast cancer: early menarche (<10 years) or late menopause (after the age of 50 years); nulliparity or late 1st pregnancy; personal history of cancer of the other breast (and proliferative forms of fibrocystic disease) or endometrium; and history of breast cancer in mother or sister (especially if bilateral and/or premenopausal).

9) c.

Objective: review factors that affect serum creatinine measurement.

Cimetidine, aspirin, and trimethoprim inhibit tubular excretion of creatinine and *increase* its serum level. Ketoacids, cephalothin, and fluocytocine are "non-creatinine chromogenes" which *increase* serum creatinine value. In liver disease, reduction of hepatic creatine formation occurs with secondary lowering of its serum concentration. Elderly people and emaciated patients have reduced muscle mass; therefore, less creatinine will be formed every day.

10) b.

Objective: review types of atrial septal defects.

ASD secundum is the commonest type, comprising about 70% of all cases. Small secundum defects are compatible with normal life and the risk of infective endocarditis is low. Complete or incomplete right bundle branch block is noticed in primum and secundum ASD. The typical plain X-ray chest film shows small aortic knob, large pulmonary artery, pulmonary plethora, and enlarged right atrium and ventricle.

11) d.

Objective: differentiate between mild, moderate, severe, and life-threatening asthmatic attacks.

Bradycardia is usually "relative", and pulsus paradoxus is usually absent. The patient is unable to speak a word and it is impossible for him to complete a whole sentence. Patients may demonstrate paradoxical thoraco-abdominal movements. The SaO_2 is typically <90%. Very little air enters into the lungs with each breath; this renders the chest silent (no wheeze would be heard). The PaO_2 is <60 mmHg, $PaCO_2$ >45 mm Hg, and PEFR <50% of predicted value (or patient's previous best value).

12) d.

Objective: review antidotes in intoxications.

Atropine and pralidoxime are used in organophosphorus poisoning. Flumazenil is used in benzodiazepine poisoning while nalaxone is effective in reversing opioids intoxication. Sodium nitrate or sodium thiosulphate are antidotes of cyanide poisoning; fomepizole (4-methylpyrazol) is helpful in ethylene glycol and methanol poisonings. Hyperbaric O_2 is a mode therapy for Carbone monoxide poisoning. N-acetylcysteine is the antidote of choice for paracetamol poisoning. Isoniazid toxicity can be treated with vitamin B_6.

13) d.

Objective: review hypersensitivity reactions.

Serum sickness is an immune-complex disease (type III hypersensitivity reaction) that typically develops 7-10 days after antigenic exposure; antitoxin, antisera, and medications (allopurinol, sulphonamides, cephalosporins, penicillins, phenytoin, procainamide, and ibuprofen). Fever, arthritis, and nephritis may also develop. The culprit antibody is an IgG antibody. The disease is usually self-limiting, but corticosteroids may be used in severe cases. Aspirin can relieve join pains.

14) e.

Objective: review hereditary causes of hyperbilirubinemia.

Gilbert's syndrome results from glucuronyl transferase deficiency; this would result in indirect hyperbilirubinemia with otherwise normal liver function tests. Prolonged fasting, for 24-36 hours, induces the appearance of the jaundice. The prognosis is excellent and the life span is normal. Reassurance is very important.

15) c.
Objective: review anxiety disorders.

Obsessive-compulsive disorder (OCD) has a prevalence of 2-3% of the general population; it is not a rare disease. About 70% of patients have coexistent major depression with a risk of suicide. There is overlapping with nail biting as well as hair pulling, tics, Tourette's syndrome, hypochondriasis, and eating disorders; so-called OCD spectrum. Most patients are intelligent, orderly, and conscientious, and deny their symptoms. Patients must be asked specifically about their symptoms. Risk factors for OCD are those of stressful life, including young age, unemployment, separated or divorced.

16) a.

Objective: review acute intermittent porphyria.

Acute intermittent porphyria is caused by partial deficiency of porphobilinogen deaminase a resulting in accumulation of δ-aminolevulinic acid and prophobilinogen in urine. The severe lancinating abdominal pain is neurogenic in origin; there are no fever or leukocytosis. The pain may be very severe, wrongly necessitating laparotomy. Hyponatremia is common and is usually profound; it is due to syndrome of inappropriate secretion of ADH. The freshly voided urine looks normal; it turns dark upon standing in air and light for some time. Intravenous hematin and glucose water infusion are treatment options during acute attacks. .

17) c.

Objective: review factors that interfere with HbA1c assay.

Thalassemia major lowers HbA1c; the high level of hemoglobin F is not identified by the immunoassay. High blood levels of acetylated hemoglobin (e.g., aspirin therapy) make cation exchange chromatography over-estimate HbA1c.

Increased rate of hemoglobin carbamylation in uremic patients makes cation exchange chromatography over-estimate HbA1c. Hemolysis, blood loss, or any factor that lessens red cell survival or life span, will falsely lower HbA1c.

18) e.

Objective: review celiac disease.

Celiac disease is common in northern Europeans; it is rare in Asians and Africans. The HLA DQ2 and HLA DQ8 haplotypes are often associated with celiac disease. The overall clinical picture depends on the length of the bowel involved, not the severity of local wall dysfunction. Serum IgA anti-gliadin is useful to monitor treatment compliance; its titer usually falls within 3-6 months of starting gluten-free diet. Small bowel T-cell lymphoma occurs in about 10% of patients who don't comply with their gluten-free diet.

19) c.

Objective: review AIDS-defining illnesses.

The list of AIDS-defining illnesses which don't require lab evidence of HIV infection is long. It includes: candidiasis of the bronchi, lungs, or esophagus; progressive multifocal encephalopathy; Kaposi's sarcoma in patients <60 years; and primary CNS lymphoma in patients <60 years. Recurrent non-typhoidal Salmonella bacteremia and recurrent bacterial pneumonia require a *lab evidence* of HIV infection.

20) a.

Objective: review medications and diseases which affect serum cholesterol.

Malignancy, thyrotoxicosis, and liver cirrhosis lower serum levels of total cholesterol. Beta blockers, diuretics, nephrotic syndrome, chronic alcohol ingestion, and chronic cholestasis increase serum total cholesterol. Note that alcohol also increases serum triglyceride and HDL cholesterol.

21) e.

Objective: review management of secondary prophylaxis of ischemic heart disease.

Statins, ACE inhibitors, beta blockers, and aspirin are the medications of secondary prophylaxis against ischemic cardiac events. Nitrates provide symptomatic improvement only, and they are not part of this prophylaxis.

22) e.

Objective: review Buerger's disease.

In Buerger's disease, arteries and veins are involved with inflammation and thrombosis. It typically attacks young smoker males. Characteristically, the proximal pulses are intact while distal pulses are absent or feeble. The disease is never confined to one limb; however, other limbs may be asymptomatic. In 75% of cases, skin ulcers develop around nails.

23) b.

Objective: review the diagnostic criteria of Allergic bronchopulmonary Aspergillosis.

In allergic bronchopulmonary aspergillosis, the total and specific serum IgE is characteristically raised (sometimes reaching very high levels), and there is peripheral blood eosinophilia. The pulmonary opacities are typically fleeting; however, they are sometimes fixed. The resulting bronchiectasis is a central (proximal) one. Skin response to *Aspergillus* antigen is strongly positive, while serum precipitin testing is weakly positive; it is a non-invasive pathology.

24) a.

Objective: review drug-induced cardiac toxicity.

Cardiac dysfunction may also result from chlorpromazine, chloroquine, emetine, and arsenicals. Candesartan is an angiotensin receptor blocker; it is used in the treatment of heart failure.

25) b.

Objective: review the causes of hypocalcemia.

Hypocalcemia may occur in chronic alcoholics, malabsorption, hyperphosphatemia (secondary fall in serum calcium), and short bowel syndrome (and bypass). Milk-alkali syndrome causes hypercalcemia. Nephrotic syndrome results in hypoalbuminemia.

Hypoalbuminemia reduces total serum calcium, but the ionized form remains unaffected. In clinical practice, hypoalbuminemia is the commonest cause of hypocalcemia.

26) c.

Objective: review polycythemias.

In polycythemia vera, the red cell mass increases (as does the hematocrit). The ESR is characteristically very low. Serum vitamin B_{12} is elevated (due to increased production and release of transcobalamine III by white cells). Enlarged spleen is one of the major criteria for the diagnosis. Erythropoietin is suppressed and its urinary levels are very low.

27) a.

Objective: review renal stones.

Hypocitraturia results from chronic diarrhea and distal renal tubular acidosis; citrate binds calcium and prevents stone formation. Therefore, hypocitraturia is a risk factor for renal stone formation. Another risk factor is hyperoxaluria (usually due to primary small bowel disorders). Hypercalciuria of > 200 mg/day confers an increased risk of nephrolithiasis. Hyperuricosuria can result in calcium and/or uric acid ones. Urinary tract obstruction and recurrent urinary tract infections predispose to struvite stones formation.

28) b.

Objective: review acromegaly.

Macroglossia combined with hypertrophy of pharyngeal and laryngeal tissues increases the incidence of obstructive sleep apnea in acromegaly patients. Excessive sweating and sebum production are signs of disease activity. Kyphosis and arthropathy affect many patients. Hypertension and diabetes mellitus are both common.

29) e

Objective: review fibromyalgia.

The ESR and white cell counts are normal in fibromyalgia. The joints look normal and there are no swelling, redness, or warmness. Headache and numbness sensation in the body are common accompaniments. Graded exercise program combined with a tricyclic antidepressant are the usual treatment.

30) b.

Objective: review pseudotumor cerebri.

The intracranial pressure is elevated but it may fluctuate widely, but never returns to its normal range. Abducens palsy is common and is a false-localizing sign; it could be unilateral or bilateral. In idiopathic pseudotumor cerebri, all lab parameters should be normal. Headache is the commonest presentation.

31) c.

Objective: review tamoxifen.

Tamoxifen is an estrogen-receptor modulator agent, having antagonistic and partial agonistic activities. Tamoxifen is currently used for the treatment of both early and advanced estrogen receptor-positive breast cancer in premenopausal and post-menopausal women. Besides its use in breast cancer, it is being increasingly used to reduce bone fractures. With prolonged tamoxifen therapy, there is a small age-dependent, but significant, increase in the risk of endometrial cancer (including sarcomas). It reduces the risk of developing cancer in the other breast by about 40%. It exerts a favorable effect on lipid profile because of its partial estrogen agonistic activity.

32) e.

Objective: review major depression.

Depressed patients may have weight gain or loss. Insomnia may be early (difficulty falling asleep) or late (the patient awakes after few hours and cannot sleep again). The nadir of the depressed mood is mainly seen in the early morning. Many patients report poor concentration; testing for memory may be abnormal resulting in pseudo-dementia. Agitation or physical retardation occurs. The development of psychosis is a marker of severe disease. Risk of homicide is seen in schizophrenia. Depressed patients are at risk of committing suicide.

33) b.

Objective: review inflammatory bowel disease.

Ulcerative colitis is more common in non-smoker and ex-smokers; Crohn's disease patients are usually smokers. The rectum is always involved early in ulcerative colitis. Due to incompetence of the ileocecal valve, backwash ileitis may occur. Appendectomy before the age of 20 years is protective against the future development of ulcerative colitis; the mechanism is unknown. Surprisingly, ulcerative colitis severity is lower in active smokers; actually, it may worsen in those who quit smoking!

34) d.

Objective: review tinea versicolor.

The fine skin scales of tinea versicolor are usually not visible; they are usually observed after skin scraping. The involved area has hypo-pigmented macules which characteristically do not tan; a hyper-pigmented form is rare. The central part of the upper body is the usual target. The lesions are asymptomatic. In the presence of larger periorificial lesions, vitiligo becomes a differential diagnosis.

35) c.

Objective: review selective IgA deficiency.

In this disease, serum IgM is normal; rarely, selective IgG subclass 2 deficiency may coexist. Atopy and celiac disease are the other associations. When IgG subclass 2 deficiency is present, the patient becomes more susceptible to infection with encapsulated bacteria and the degree of immune deficiency increases. Similar to common variable immune deficiency syndrome, autoimmune diseases are common. Transfusion of blood (or blood products) is dangerous because of the presence of anti-IgA antibodies (which can result in anaphylaxis or serum sickness).

36) b.

Objective: review the effects of drugs and medications on vitamin absorption and metabolism.

Aspirin increases urinary ascorbic acid excretion and lowers serum levels of this vitamin.

Salphasalazine and cholestyramine interfere with folate absorption. Prolonged therapy with H_2-blockers decreases iron and vitamin B_{12} absorption. Vitamin D and K turnovers are accelerated by the use of phenytoin. Diuretics enhance the urinary loss of vitamin B_1 and B_6.

37) c.

Objective: review cardiovascular changes in pregnancy.

The stroke volume, ejection fraction, and cardiac output increase during pregnancy. The diastolic blood pressure falls significantly because of the reduction in the peripheral vascular resistance (resulting from vasodilatation and blood flow through low resistant shunt flow of the placenta). The systolic blood pressure remains unchanged or rises slightly. The resting heart rate increases, mainly in the 3rd trimester. A third heart sound is common, as is the presence of pulmonary flow murmur. Due to the more horizontal position of the heart, some Q waves may appear in the inferior 12-lead ECG.

38) e.

Objective: review central and branch retinal artery occlusions.

Most patients present with sudden visual loss, usually to finger counting or less. The retina is diffusely pale (milky-white). During the first few days, a cherry-red macular spot appears. Optic atrophy is a common aftermath, and the retinal arteries usually remain attenuated. Upon recovery, the visual field is usually restricted to an island of vision in the temporal field.

39) a.

Objective: review risk factors for obstructive sleep apnea.

Hypothyroidism, obesity, acromegaly, receding chin, large tongue, tonsillar hypertrophy, and cigarette smoking are risk factors for developing obstructive sleep apnea.

40) e.

Objective: review hairy cell leukemia.

Hairy cell leukemia is an indolent B-cell malignancy. Anemia is almost universal, and at least 75% of patients have leucopenia and thrombocytopenia. Male to female ratio is 5:1. Monocyotpenia is characteristic; there is increased susceptibility to tuberculous infections. The bone marrow aspiration is difficult; a "dry tap".

41) d.

Objective: review chest X-ray findings in lung cancer.

Some bronchogenic cancers are entirely endobronchial; the plain chest X-ray film cannot reveal them. Small and early tumors may also have a normal plain film. Lung, lobar, or segmental collapse may occur. Phrenic nerve involvement would cause elevation of the paralyzed hemi-diaphragm. One or both hili may enlarge because of the tumor itself or enlarged lymph nodes. Persistent infiltrates on plain chest films should always be taken seriously.

42) e.

Objective: review nephrotic syndromes.

Focal segmental glomerulosclerosis has been linked to morbid obesity, HIV infection, and heroin abuse. The renal function is impaired in 50% of cases at the time of diagnosis; this usually ends-up with chronic renal failure within 6-8 years. The resulting proteinuria is non-selective. Overall, the prognosis is poor and steroids are usually ineffective.

43) c.

Objective: review toxic shock syndrome.

Toxic shock syndrome is an exotoxin-mediated disease that results from infection with certain *S. aureus* strains; therefore, bacteremia is not a factor, and blood cultures are negative. The bacteria release enterotoxin type B or TSST-1. High fever is common at onset. Hypotension may be profound and renal and heart failures may develop. Skin desquamation, especially of the palms and soles, is typically seen during recovery. A similar syndrome may result from certain strains of streptococci; so-called toxic shock-like syndrome (TSLS) or streptococcal toxic shock syndrome (STSS). This bacteria releases streptococcal pyrogenic exotoxins.

44) c.

Objective: review Kawasaki's disease.

Kawasaki disease is a disease of children, (mainly affecting those under the age of 5 years. Lip fissuring, red mouth, and conjunctival hyperemia are one of the core features of the disease. The enlarged cervical lymph nodes are non-suppurative. Coronary artery aneurysmal formation may develop, usually transiently. Aspirin and intravenous immunoglobulin are the mainstay treatments.

45) d.

Objective: review pheochromocytoma.

In 60% of cases, the blood pressure is sustained (rather than labile). With dopamine-secreting tumors, postural hypotension occurs. Thyroid function is intact. Weight loss, constipation, impaired glucose tolerance test, and nausea are common. Abdominal pain mainly occurs during attacks.

46) e.

Objective: review the complications and management of acute pancreatitis.

The development of adult respiratory distress syndrome confers a high mortality in acute pancreatitis; this usually happens 3-7 days after onset of pancreatitis. Cardiac failure may superimpose on the picture. Pre-renal failure may occur secondary to volume depletion. Erosive gastritis or erosion of the duodenum or colon may result in GIT hemorrhage. Splenic vein thrombosis is a well-recognized complication causing rapid enlargement of the spleen. Transient hyperglycemia usually develops; however, permanent diabetes rarely follows.

47) c.

Objective: review the management of acute digoxin poisoning.

Nausea, vomiting, and yellow vision are features of acute digoxin intoxication. This poisoning causes hyperkalemia; note that hypokalemia precipitates toxicity. On the other hand, *chronic* toxicity is often accompanied by hypokalemia and hypomagnesemia.

Bradycardia and AV blocks are very common. In adults ventricular ectopics are common. Non-paroxysmal atrial tachycardia with heart block and bidirectional ventricular tachycardia are particularly characteristic of severe digitalis toxicity. Note that digoxin toxicity may cause almost any dysrhythmia.

48) a.

Objective: review features of fragile X-mental retardation syndrome.

Macroorchidism is mainly seen after puberty. There are large protruding ears and long face. The jaw is long. Mitral valve prolapse is common. Most patients have a high pitched voice. There are hyper-extensible finger joints and hyper-extensible (double jointed) thumb. The skin is soft and there is generalized hypotonia. A wide range of behavioral abnormalities can be seen; actually, some individuals with fragile X syndrome also meet the diagnostic criteria for autism.

49) c.

Objective: review the management of hypertrophic cardiomyopathy.

Digoxin is contraindicated; it increases the left ventricular outflow tract obstruction. Hydralazine is a vasodilator and produces the same effect of digoxin. Beta blockers are the first-line agents, especially in high risk patients; disopyramide and calcium channel blockers are second line agents. Diuretics can be used judiciously.

50) b.

Objective: review the role of radiotherapy in treating cancers.

Laryngeal squamous cell cancer can be treated by radiotherapy, permitting cure without voice loss; the same intent applies to cancers of the pharynx and oral cavity. Breast and rectal cancers cannot be cured by radiotherapy. Radiotherapy may cure Hodgkin's disease patients, especially stage IA disease. Cancers of the prostate, vagina, and cervix may at times cured by radiotherapy.

51) c.

Objective: review the pulmonary side effect profile of medications.

This patient most likely takes amiodarone, which can result in interstitial fibrosis. The fibrosis is mainly bibasal. Amiodarone has acute/subacute/chronic pulmonary toxicities.

52) a.

Objective: differentiate between idiopathic Parkinson's disease and Parkinsonian-plus syndromes.

The presence of extensor planter reflex is highly unusual in idiopathic Parkinson's disease, as is the presence of spasticity, exaggerated reflexes, cerebellar signs, and muscle atrophy. These atypical features should prompt the physician to do brain imaging.

53) c.

Objective: review the chromosomes involved in various genetic diseases.

The mutated 508 locus in the CFTR gene (the commonest mutation in cystic fibrosis) lies on chromosome 7.

54) d.

Objective: review Lyme's disease.

The patient's dropped beats indicate heart block. He most likely has developed Lyme's disease. The initial tick bite might not be recalled by the patient and careful history taking is very important. Many healthy individuals living in endemic areas have positive serology for Lyme's disease, however.

55) d.

Objective: certain HIV diseases have a CD4+ count threshold of appearance.

At least 90% of *P. carinii* pneumonia cases appear when the CD4+ cell count falls below 200 cells/mm³.

56) b.

Objective: review the normal cellular constituents of the BAL fluid.

Even if you have no idea about this topic, you must have remembered that the inner side of normal alveoli is patrolled by "alveolar macrophages". Neutrophilic BAL is highly abnormal and implies a poor prognosis in many diseases, such as idiopathic pulmonary fibrosis. Eosinophilic BAL reflects pulmonary eosinophilia.

57) d.

Objective: review the indications of treatment in sarcoidosis.

Obviously, the patient has cardiac involvement in the form of AV block; this calls for steroid treatment. Besides, lupus pernio necessitates steroids administration, as it indicates a high likelihood of progressing to chronic fibrosing sarcoidosis. Meanwhile, a consideration for permanent pacemaker should be taken into account, but it is not the urgent choice in this man.

58) d.

Objective: review renal stone risk factors and types.

Urinary tract infections and obstruction are strongly associated with future development of magnesium ammonium phosphate (struvite) stones because of urea-splitting microorganisms.

59) e.

Objective: review the causes of vitamin B_{12} deficiency.

Vitamin B_{12} stores are sufficient for 3-5 years only. The terminal ileum is the site of vitamin B_{12} absorption, which was surgically removed in this patient.

60) e.

Objective: review drug-induced psychosis.

The patient has features of psychosis, which may be due to schizophrenia or other diseases. Drug abuse is common among high school students, which is by far, more common than schizophrenia as a cause of psychosis, personality changes, and aggression in this population.

61) c.

Objective: review the markers of activity in SLE.

Plasma C3, C4, CH50, and anti-dsDNA are useful markers of disease activity. ANA is used in screening, and has nothing to do with disease activity.

62) b.

Objective: review causes of hyperprolactinemia and their approximate prolactin value.

The serum prolactin is between 1000-2000 u/L. Given the size of the lesion, it is most likely that represents disconnection hyperprolactinemia (because of damage to the pituitary stalk) resulting from non-functioning pituitary adenoma. A macroprolactinoma would produce a very higher serum value.

63) a.

Objective: review the management of upper GIT bleeding.

Whatever the cause is, the first step is to resuscitate the patient. The patient should be stabilized before doing any further investigations.

64) a.

Objective: differentiate between pseudohypoparathyroidism and hypoparathyroidism.

This short scenario fits pseudohypoparathyroidism; 4th and 5th short metacarpals will be revealed by doing X-ray film of the hand.

65) d.

Objective: review the management of SVT.

This otherwise healthy patient is hemodynamically stable, and a trial of carotid massage (or any other maneuver that increases the vagal tone) may be successful in terminating the attack of supraventricular tachycardia. Going directly to pharmacological therapy would be unreasonable, and DC shock is not indicated in hemodynamically stable patients.

66) d.

Objective: review headache syndromes.

The history is highly suggestive of cluster headache. These attacks respond to high dose oral prednisolone, triptans, or high flow, high concentration nasal O_2.

67) b.

Objective: review colorectal cancers.

Apple-core sign points out towards colonic cancer. This appearance usually carries a poor prognosis.

68) d.

Objective: review the lab findings and diagnostic criteria of anorexia nervosa.

In anorexia nervosa, amenorrhea is one of the core features; the presence of normal menstrual cycle strongly throws a doubt on the diagnosis. Growth hormone rises because of stress and starvation.

69) b.

Objective: review the causes of membranous nephropathy.

The clinical picture is suggestive of a body water retaining pathology. The recent change in the character of the cough may well indicate the development of lung cancer. The foamy urine reflects high urinary protein content. The patient has lung cancer-associated membranous nephropathy.

70) e.

Objective: review lithium toxicity.

The patient most likely takes daily lithium therapy. The commonest predisposing conditions for lithium toxicity are diarrhea and the addition of a thiazide diuretic.

71) c.

Objective: review the complications of cirrhosis.

This patient seems to have alcoholic cirrhosis that is complicated by spontaneous bacterial peritonitis (SBP). The latter has precipitated hepatic encephalopathy. SBP is the presenting problem in this patient; hepatic encephalopathy is secondary to liver decompensation.

72) d.

Objective: review subclinical hyperthyroidism.

The serum free T_4 is high normal and there is suppressed serum TSH; this is subclinical hyperthyroidism. This category has been shown to confer a risk of atrial fibrillation and osteoporosis.

73) b.

Objective: review the causes of isolated thrombocytopenia.

Acute onset of generalized bleeding tendency in children with low platelets on blood counts (and normal hemoglobin and white cells) is highly suggestive of idiopathic thrombocytopenic purpura.

74) a.

Objective: differentiate upper lung zone fibrosis from lower zones fibrosis.

Generally, idiopathic pulmonary fibrosis, drug-induced pulmonary fibrosis, asbestosis, and connective tissue diseases are the causes of predominantly *lower* zones fibrosis.

75) a.

Objective: review factors that exacerbate psoriasis.

Patients usually improve on sun exposure; this is the rational of using UV light as a mode of therapy in psoriasis.

76) e.

Objective: review the genetics of inherited immune deficiency states.

The inheritance of common variable immune deficiency syndrome is still unknown, and the risk of passing the disease to the next generation is zero, i.e., it is sporadic.

77) d.

Objective: review the HLA association of various diseases.

Hereditary hemochromatosis has been associated with HLA A3.

78) c.

Objective: review NNT, ARR, and RRR.

Questions addressing NNT (number needed to treat), absolute risk reduction (ARR), and relative risk reduction (RRR) are commonly encountered in examinations. Absolute risk reduction is calculated as ARR=treated group – control group; the "medisan" study has resulted in 4% ARR (i.e., 10 minus 6). Relative risk reduction (RRR) is measured as RRR= (treated group-control group)/treated group; our study has achieved a 40% RRR in seizure occurrence. The NNT is obtained by dividing 1 over ARR (i.e., 1/4); you had to treat 25 head trauma patients in order to prevent one patient from developing seizures.

79) d.

Objective: review the management of complications resulting from acute subarachnoid hemorrhage.

Only nimodipine has been shown to reduce the incidence of cerebral vasospasm after subarachnoid hemorrhage. Papaverine is used in vasospasm treatment, not in the prophylaxis.

80) b.

Objective: review tumor markers in testicular malignancies. .

While β-hCG is secreted by seminomatous as well as non-seminomatous testicular tumors, only non-seminomatous or mixed testicular tumors secrete alpha fetoprotein. This patient probably has a mixed tumor type.

81) c.

Objective: review causes of acute hepatic failure.

Paracetamol poisoning ranks first on the list of causes of acute hepatic failure in the Western World. Viral hepatitides are the usual causes in the developing countries.

82) d.

Objective: review substances that affect serum uric acid level.

Cyclosporine, low-dose aspirin, pyrazinamide, frusemide, thiazides, and alcohol reduce urinary uric acid excretion.

83) b.

Objective: review skin tumors.

The description is that of basal cell carcinoma.

84) c.

Objective: review the causes of epistaxis.

Osler-Weber-Rendu, nasal tumors, and coagulopathy are uncommon causes of epistaxis. Epistaxis is a common, and most cases are benign and self-limiting.

85) b.

Objective: review the evaluation of thyroid diseases.

Serum T3 is normal is 25% of cases of overt hypothyroidism (because of compensated peripheral mechanisms). Serum TSH is the most sensitive indicator of primary thyroid dysfunction; it is misleading in the secondary forms, however. Serum prolactin may be raised in hypothyroidism, but it is not diagnostic of it.

86) c.

Objective: review the management of atopic dermatitis.

Topical emollients are the first-line agents in the treatment of atopic dermatitis. Topical steroids should be used sparingly and cautiously. Topical vitamin D has been shown to be effective in chronic plaque psoriasis.

87) d.

Objective: review the clinical features of polymyositis and compare them with those of dermatomyositis.

She has tender proximal myopathy as well as interstitial pulmonary fibrosis. Note the exertional dyspnea; positive serum anti-Jo1 antibodies correlate well with parenchymal lung involvement. She has also cardiac involvement (heart block; dropped beats). Her dyspnea might well be due to involvement of the heart and lung. Skin manifestations of dermatomyositis are absent.

88) c.

Objective: review the management of adult onset epilepsy.

This man has 3 unprovoked attacks of generalized tonic-clonic convulsions within 1 week. He should receive an anti-epileptic medication while the investigations are being carried out (as his risk of recurrent seizures is high). Echocardiography has no justification; nothing in the history suggests a cardiac cause (infective endocarditis or cardiac myxoma with cerebral embolism). EEG should not be used as the sole diagnostic tool; it *confirms* the clinical suspicion and may show a focal epileptic focus. The patient's seizures might be partial seizures with secondary generalization; the very brief focal fit may pass unnoticed. Brain CT scan with a contrast would look for tumors as the cause of this adult onset epilepsy.

89) d.

Objective: review the lab features of iron deficiency anemia.

This patient most likely has peri-menopausal dysfunctional uterine bleeding; this would result in iron deficiency anemia (IDA). Thrombocytosis is a well-known consequence of long-standing IDA.

The white cell count remains unaffected. The iron biding capacity is increased, attempting to increase the marrow delivery of iron.

90) d.

Objective: review the new treatments of chronic hepatitis B infections.

Pegylated interferon, lamivudine, entecavir, and adefovir are useful in the treatment of chronic hepatitis B viral infections. The latter 2 are useful in treating lamivudine-resistant cases. The objective of treatment is to lower serum transaminases, induce seroconversion towards the development of anti-HBe antibodies, and to lower hepatitis B viral DNA level in blood; complete elimination of the HBs antigen is rarely achieved, however.

81) b.

Objective: review the imaging studies in acute aortic dissections.

The patient's presentation is highly suggestive of acute aortic dissection. Plain chest X-ray films are neither sensitive nor specific for aortic dissections. The ECG may show inferior wall myocardial ischemia or infarction if the osteum of the right coronary artery is involved in type A dissections.

92) a.

Objective: review the defective site in immune-deficiency states.

X-linked Bruton's agammaglobulinemia has no detectable B cells in the peripheral blood. This is due to maturation arrest at the pre-B cell stage in the bone marrow; the core defect is in the *BTK* gene on chromosome X.

93) b.

Objective: review the management of lymphedema.

Diuretics poorly, if ever, mobilize the lymphedematous fluid; besides, they contract the intravascular volume. Therefore, they should be avoided.

94) d.

Objective: differentiate between typical and atypical pneumonia.

The overall picture is suggestive of atypical pneumonia. The positive serum cold agglutinin test points out towards *M. pneumonae*.

95) c.

Objective: review signs of severity in valvular lesions.

Severe mitral regurgitation is reflected by the presence of dilated thrusting cardiac apex, long duration of the holosystolic murmur, presence of third heart sound, presence of mid-diastolic flow murmur, and signs of heart failure (bibasal crackles, leg edema, raised JVP,…etc.).

96) b.

Objective: review the general anatomy of the ventricular system of the brain.

The lateral ventricles communicate inferiorly with the 3rd ventricle through the foramen of Monroe. Then, the CSF flows via the cerebral aqueduct into the 4th ventricle. Finally, the CSF escapes through one median and 2 lateral foramina into the subarachnoid space.

97) d.

Objective: review the guidelines for interpreting tuberculin skin testing.

A tuberculin skin test with an induration between 5-9 mm is considered positive in the following groups of individuals: close contacts with smear positive pulmonary tuberculosis patient; immune-compromised patients; recipients of solid organ transplants; those who have abnormal chest X-rays suggestive of old or current tuberculosis; and HIV patients.

98) e.

Objective: review the extra-renal manifestations of adult polycystic kidney disease.

Mitral valve prolapse, mitral regurgitation, aortic regurgitation, aortic aneurysms, pancreatic and hepatic cysts, and divarication of recti may be encountered in patients with adult polycystic kidneys disease.

99) a.

Objective: review mixed acid/base disorders.

Gram-negative sepsis stimulates the respiratory center directly (with resultant rapid shallow respiration) and produces lactic acidosis; this is by far the commonest cause of combined metabolic acidosis and respiratory alkalosis.

100) c.

Objective: review vaccination of HIV patients.

BCG and oral polio vaccines are contraindicated in HIV-infected patients. MMR vaccine should be given to HIV-infected patients, except those who are severely immune compromised. The varicella vaccine is currently not recommended for HIV-infected adults or children, but its safety is being evaluated in HIV-infected adults. HIV-infected travelers with CD4+ counts over 200 cells/mm^3 may be safely immunized with the yellow fever vaccine; however, efficacy cannot be guaranteed since only 35% of patients will be seroconverted following vaccine administration. Since the efficacy of BCG vaccine in HIV-infected patients is unknown and there is a risk of disseminated disease, it is recommended that BCG vaccine is not to be used in those patients, even if the risk of acquiring tuberculosis is high.

Further reading:

You may also try other MRCP self-assessment books, which were written by Osama S. M. Amin:

1. Get Through MRCP;BOFs.

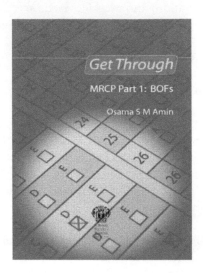

2. Self-Assessment for MRCP(UK) part I.

3. Neurology: Self-Assessment for MRCP(UK) and MRCP(I).

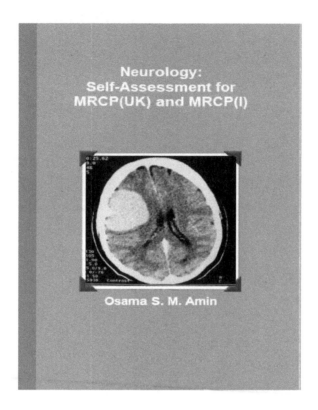

This page was intentionally left blank

CPSIA information can be obtained
at www.ICGtesting.com
Printed in the USA
LVHW03s1729120718
583537LV00003B/509/P

9 781365 585159